Moritz Knorr

Self-Calibration of Multi-Camera Systems for Vehicle Surround Sensing

Schriftenreihe
Institut für Mess- und Regelungstechnik,
Karlsruher Institut für Technologie (KIT)

Band 041

Eine Übersicht aller bisher in dieser Schriftenreihe erschienenen Bände finden Sie am Ende des Buchs.

Self-Calibration of Multi-Camera Systems for Vehicle Surround Sensing

by
Moritz Knorr

KIT Scientific Publishing

Dissertation, Karlsruher Institut für Technologie
KIT-Fakultät für Maschinenbau

Tag der mündlichen Prüfung: 25. April 2017
Referenten: Prof. Dr.-Ing. C. Stiller, Prof. Dr.-Ing. S. Hinz

Impressum

KIT Scientific Publishing

Karlsruher Institut für Technologie (KIT)
KIT Scientific Publishing
Straße am Forum 2
D-76131 Karlsruhe

KIT Scientific Publishing is a registered trademark
of Karlsruhe Institute of Technology.
Reprint using the book cover is not allowed.

www.ksp.kit.edu

Print on Demand 2018 – Gedruckt auf FSC-zertifiziertem Papier

ISSN 1613-4214
ISBN 978-3-7315-0765-9
DOI 10.5445/KSP/1000080527

Vorwort

Die vorliegende Arbeit entstand während meiner Doktorandentätigkeit bei der Robert Bosch GmbH in Hildesheim und wurde von Prof. Dr.-Ing. Christoph Stiller vom Institut für Mess- und Regelungstechnik des Karlsruher Instituts für Technologie (KIT) betreut. Ihm gebührt mein außerordentlicher Dank für zahlreiche konstruktive Gespräche und die umfassende Betreuung, sowie die Möglichkeit, mich mit den wissenschaftlichen Mitarbeitern des Instituts zu vernetzen. Prof. Dr.-Ing. Stefan Hinz danke ich ganz herzlich für die Übernahme des Korreferats und für seine hilfreichen Anregungen zur Arbeit.
Bedanken möchte ich mich darüber hinaus ganz besonders bei Wolfgang Niehsen für die inhaltliche Betreuung seitens Bosch und viele fruchtbare Diskussionen. Ihm, sowie Stephan Simon möchte ich im Speziellen dafür danken, mich stets motiviert zu haben, viele Probleme bis ins Detail zu ergründen und zu untersuchen. Henning von Zitzewitz danke ich für viele wertvolle Diskussionen und Anregungen zum Thema Geometrie und 3D-Konstruktion.
Diese Arbeit wäre nicht möglich gewesen, ohne die großzügige Unterstützung durch viele meiner Kollegen. Besonders bedanken möchte ich mich an dieser Stelle auch bei Sebastian Lauer für die Hilfe bei der Ausrüstung des Versuchsträgers, bei Dimitrios Bariamis für viele Hinweise und förderliche Gespräche, sowie bei Thomas Wenzel für die mühevolle Arbeit des Korrekturlesens.
Die Anfertigung dieser Arbeit hat sich insgesamt über einen längeren Zeitraum erstreckt, als ich mir eingestehen will. Ich möchte mich daher noch einmal ganz herzlich für das Verständnis und die Unterstützung durch meine Freunde und meine Freundin bedanken. Im Besonderen gilt dies auch für meine Mutter und meine Tante, denen ich diese Arbeit widmen möchte.

Hildesheim, im Februar 2018 *Moritz Knorr*

Kurzfassung

Multikamerasysteme werden heute bereits in einer Vielzahl von Fahrzeugen und mobilen Robotern eingesetzt. Die Anwendungen reichen dabei von einfachen Assistenzfunktionen wie der Erzeugung einer virtuellen Rundumsicht bis hin zur Umfelderfassung, wie sie für teil- und vollautomatisches Fahren benötigt wird. Damit aus den Kamerabildern metrische Größen wie Distanzen und Winkel abgeleitet werden können und ein konsistentes Umfeldmodell aufgebaut werden kann, muss das Abbildungsverhalten der einzelnen Kameras sowie deren relative Lage zueinander bekannt sein.

Insbesondere die Bestimmung der relativen Lage der Kameras zueinander, die durch die extrinsische Kalibrierung beschrieben wird, ist aufwendig, da sie nur im Gesamtverbund erfolgen kann. Darüber hinaus ist zu erwarten, dass es über die Lebensdauer des Fahrzeugs hinweg zu nicht vernachlässigbaren Veränderungen durch äußere Einflüsse kommt. Um den hohen Zeit- und Kostenaufwand einer regelmäßigen Wartung zu vermeiden, ist ein Selbstkalibrierungsverfahren erforderlich, das die extrinsischen Kalibrierparameter fortlaufend nachschätzt.

Für die Selbstkalibrierung wird typischerweise das Vorhandensein überlappender Sichtbereiche ausgenutzt, um die extrinsische Kalibrierung auf der Basis von Bildkorrespondenzen zu schätzen. Falls die Sichtbereiche mehrerer Kameras jedoch nicht überlappen, lassen sich die Kalibrierparameter auch aus den relativen Bewegungen ableiten, die die einzelnen Kameras beobachten. Die Bewegung typischer Straßenfahrzeuge lässt dabei jedoch nicht die Bestimmung aller Kalibrierparameter zu. Um die vollständige Schätzung der Parameter zu ermöglichen, lassen sich weitere Bedingungsgleichungen, die sich z.B. aus der Beobachtung der Bodenebene ergeben, einbinden. In dieser Arbeit wird dazu in einer theoretischen Analyse gezeigt, welche Parameter sich aus der Kombination verschiedener Bedingungsgleichungen eindeutig bestimmen lassen.

Um das Umfeld eines Fahrzeugs vollständig erfassen zu können, werden typischerweise Objektive, wie zum Beispiel Fischaugenobjektive, eingesetzt, die einen sehr großen Bildwinkel ermöglichen. In dieser Arbeit wird ein Verfahren zur Bestimmung von Bildkorrespondenzen vorgeschlagen, das die geometrischen Verzerrungen, die sich durch die Verwendung von Fischaugenobjektiven und sich stark ändernden Ansichten ergeben, berücksichtigt. Darauf

aufbauend stellen wir ein robustes Verfahren zur Nachführung der Parameter der Bodenebene vor.
Basierend auf der theoretischen Analyse der Beobachtbarkeit und den vorgestellten Verfahren stellen wir ein robustes, rekursives Kalibrierverfahren vor, das auf einem erweiterten Kalman-Filter aufbaut. Das vorgestellte Kalibrierverfahren zeichnet sich insbesondere durch die geringe Anzahl von internen Parametern, sowie durch die hohe Flexibilität hinsichtlich der einbezogenen Bedingungsgleichungen aus und basiert einzig auf den Bilddaten des Multikamerasystems.
In einer umfangreichen experimentellen Auswertung mit realen Daten vergleichen wir die Ergebnisse der auf unterschiedlichen Bedingungsgleichungen und Bewegungsmodellen basierenden Verfahren mit den aus einer Referenzkalibrierung bestimmten Parametern. Die besten Ergebnisse wurden dabei durch die Kombination aller vorgestellten Bedingungsgleichungen erzielt. Anhand mehrerer Beispiele zeigen wir, dass die erreichte Genauigkeit ausreichend für eine Vielzahl von Anwendungen ist.

Schlagworte: Selbstkalibrierung, Extrinsische Kalibrierung, Multikamerasystem, Umfelderfassung

Abstract

Multi-camera systems are being deployed in a variety of vehicles and mobile robots today. Applications of such systems range from driver assistance functions such as rendering a virtual panoramic view to surround sensing, which is a prerequisite for partially and fully automated driving. In order to derive metric quantities such as angles and distances from camera images and to establish a consistent representation of the vehicle environment, both, the geometric imaging characteristics of the individual cameras and the relative positions and orientations have to be known.
In particular the estimation of the relative positions and orientations, which are described by the extrinsic calibration, is troublesome since it can only be performed with the system being fully set up and since non-negligible changes of the parameters have to be expected over the life cycle of the vehicle due environmental influences. To eliminate the need for cost and labor intensive maintenance, continuous self-calibration is highly desirable.
Self-calibration typically builds upon overlapping field of views of cameras, which enables estimating the extrinsic calibration parameters using image correspondences. Motion-based calibration on the other hand does not impose constraints on the fields of view. However, the almost planar motion of typical road vehicles constitutes a special case in which only a subset of the calibration parameters can be inferred. To circumvent this problem additional constraints can be imposed, e.g. by using the ground plane as a natural reference object. In a theoretical analysis we determine the sets of parameters that can be inferred from different vehicle motion classes and camera configurations.
For visual surround sensing typically cameras with ultra-wide angle lenses, such as fisheye lenses, are employed. In order to establish image correspondences in the presence of strong geometrical distortions introduced by the lens and large viewpoint variations we propose an image warping method that exploits the knowledge about the geometric imaging process and performs a coarse scene approximation. In addition, we present a method for tracking the ground plane in the presence of structural outliers such as other planes in the scene.
Building upon the observability analysis and proposed methods we present an extended Kalman filter-based algorithm for continuous extrinsic camera self-

calibration. The filter exhibits high flexibility with regard to incorporating different measurement constraints, has a particularly low number of internal parameters, and relies solely on image data.
In an extensive evaluation we assess our algorithm quantitatively using real-world data. We compare results based on different motion models and combinations of measurement constraints against a reference calibration. It is found that the best results are obtained by combining all of the proposed measurement constraints. Using several examples we demonstrate that the achieved accuracy is sufficient for most applications.

Keywords: Self-calibration, Extrinsic Calibration, Multi-Camera System, Surround Sensing

Contents

Contents . . . . **vii**

Notation and Symbols . . . . **xi**

1 Introduction **1**
1.1 Problem Statement 4
1.2 Contribution 4
1.3 Thesis overview 5

2 Fundamentals **9**
2.1 Overlapping Fields of View 9
2.2 Motion-Based Calibration 12
2.3 Scene Constraints 14
2.4 Bayesian Filtering and Optimization 14

3 Camera Model and Two-View Geometry **17**
3.1 The Perspective Camera Model 17
3.2 The Fisheye Lens 20
3.2.1 Geometric Camera Model 20
3.2.2 Noncentrality 22
3.2.3 Light Falloff and Vignetting 24
3.3 Two-View Geometry 26
3.3.1 Camera Pose and Pose Transformation 26
3.3.2 Epipolar Geometry and the Essential Matrix 27
3.3.3 Plane Induced Homography 29

4 Extrinsic Camera Calibration **31**
4.1 Definition of the Reference Frame 32

4.2 Motion-based Calibration 34
4.2.1 Hand-Eye Calibration 35
4.2.2 The Ground Plane 36
4.2.3 Classes of Motion 37
4.2.4 Computation of Extrinsic Calibration Parameters 38
4.2.5 Summary 45
4.3 Calibration from Overlapping Fields of View 47

5 Establishing Point Correspondences **51**
5.1 Wide Baseline Matching 51
5.2 Scene Geometry Approximation 54
5.3 Image Resampling and Smoothing 55

6 Robust Homography Estimation **59**
6.1 Planar Parallax Decomposition 60
6.2 Local Adaptive Thresholds 61
6.3 Sequential Testing and Updating 67

7 Continuous Self-Calibration Based on Kalman Filtering **71**
7.1 Recursive Filtering 72
7.2 Parameterization and Motion Models 74
7.3 Extended Kalman Filter Update Stage 79
7.4 Initialization and Recovery of Vehicle Velocity 81

8 Experimental Evaluation **85**
8.1 Evaluation Dataset 85
8.2 Ground Truth and Error Metric 89
8.3 Initialization and Parameter Tuning 90
8.4 Quantitative Evaluation 91
8.4.1 Motion-Based Calibration 91
8.4.2 Overlapping Fields of View 95
8.4.3 Visual Odometry Loop Closure Error 95
8.4.4 Assessing Calibration Results at Runtime 96
8.5 Qualitative Results 105
8.5.1 Visual Odometry 105

8.5.2 Virtual Top View 107
8.5.3 Stereo Rectification 107

9 Conclusion and Future Research Directions 113

A Appendix 117
A.1 Constructing Orthonormal Matrices from Two Vectors 117
A.2 Rodrigues Formula for Rotation Matrices 117
A.3 Instantaneous Center of Rotation 118
A.4 Derivation of Equation (6.3) 118
A.5 Extended Kalman Filter 119
A.6 Sequential Processing Algorithm 120
A.7 Derivation of Equation (7.8) 122
A.8 Box Plots 123
A.9 Additional Information on Quantitative Results 124

Bibliography 125

Notation and Symbols

Acronyms

2D/3D	2/3-Dimensional
ASIFT	Affine Scale Invariant Feature Transform
BRIEF	Binary Robust Independent Elementary Features
RANSAC	Random Sample Consensus
SIFT	Scale Invariant Feature Transform
FAST	Features from Accelerated Segment Test

General Notation

Scalars	Regular, lower case: a, b, c, ...
Vectors	Bold: $\mathbf{a}$, $\mathbf{b}$, $\mathbf{X}$, ...
Matrices	Bold, upper case: $\mathbf{A}$, $\mathbf{B}$, $\mathbf{\Sigma}$, ...
Estimates	Hat operator: $\widehat{\mathbf{a}}$, $\widehat{\mathbf{A}}$, $\widehat{\mathbf{\Sigma}}$, ...
Correspondences	Prime: $\mathbf{x} \leftrightarrow \mathbf{x}'$, $\mathbf{u} \leftrightarrow \mathbf{u}'$, $\mathbf{X} \leftrightarrow \mathbf{X}'$...

Geometric Entities and Transformations

$\mathcal{C}, I$	Camera and image
$X_{\mathcal{C}}, Y_{\mathcal{C}}, Z_{\mathcal{C}}$	Camera coordinate frame axes
$\mathbf{X} = (X, Y, Z)^T$	World point
$\mathbf{x} = (x, y, z)^T$	Ray
$\mathbf{u} = (u, v)^T$	Image point
$\mathbf{l} = (l_1, l_2, l_3)^T$	Line in image coordinates
$\mathbf{C}$	Camera calibration matrix
$\boldsymbol{\kappa}(\cdot)$	Projection into the image, $\mathbf{u} = \boldsymbol{\kappa}(\mathbf{X}) = \boldsymbol{\kappa}(\mathbf{x})$

$z_0, \mathbf{u}_0 = (u_0, v_0)^T$	Focal length and Principal point
$\mathbf{H}, \mathbf{E}$	Homography and Essential matrix
$\mathbf{h}(\cdot)$	Plane induced image to image projection
$\mathbf{T}, \mathbf{R}, \mathbf{t}$	Motion induced transformation, rotation, and translation
$\mathbf{\Delta T}, \mathbf{\Delta R}, \mathbf{\Delta t}$	Rigid transformation, orientation, and displacement
r	Radius
α, θ	Off-axis angle and rotation angle
$\mathbf{r}$	Rotation axis direction (unit length)
$\mathbf{s}$	Instantaneous center of rotation
$\mathbf{n}$	Plane normal vector
$\mathbf{c} = (0, 0, c_z)^T$	Shift of the projection center
$\lambda, \boldsymbol{\lambda}$	Scale factor and vector and vector of scale factors
ω, τ	Unobservable angle and scale factor
$\boldsymbol{\Psi}(\cdot), \mathbf{J}$	Image to image mapping and Jacobian
$\mathbf{d}$	2D parallax vector
$\mathbf{A}, \mathbf{B}$	Matrices
$\mathbf{a}, \mathbf{v}$	3-vectors
$\mathbf{p}$	2-vector
ϵ	Scalar residual

Probabilistics and Kalman Filtering

$\boldsymbol{\xi}, \mathbf{P}$	State vector and covariance matrix
$\widehat{\boldsymbol{\xi}}^{-}, \widehat{\boldsymbol{\xi}}^{+}$	A priori and a posteriori state vector
$\mathbf{z}, \overline{\mathbf{z}}$	Vector of measurements and error free measurements
$m(\cdot), \mathbf{m}(\cdot), \mathbf{M}$	Scalar and vector-valued measurement constraint function and Jacobian
$\mathbf{f}(\cdot), \mathbf{F}$	State transition function and Jacobian
$\mathbf{q}, \mathbf{Q}$	Process noise vector and covariance matrix
$\mathbf{w}, \mathbf{W}$	Measurement noise and covariance matrix

$\mathbf{K}$	Kalman gain
$\mathcal{N}(\cdot,\cdot)$	Normal distribution
σ	Standard deviation
$p(;)$	Probability density function
$f(\cdot;\cdot)$	Probability density function of the noncentral χ^2 distribution
$F(\cdot;\cdot)$	Cumulative noncentral χ^2 distribution
γ	Noncentrality coefficient of noncentral χ^2 distribution
ν, η	False positive and true positive rate
ρ	Threshold

Indexing

C	Number of cameras
$c, d = \{0, \ldots, C-1\}$	Camera indeces
r	Index of the reference camera $\mathcal{C}^r$
N	Number of 2D/3D points
$i = \{0, \ldots, N-1\}$	Point index
t, k	Continuous time and discrete time index
h_k^c, r_k^c, $\mathbf{n}_k^c$	Camera height, radius, and plane normal vector in the coordinate frame of camera $\mathcal{C}^c$ at time k
$\mathbf{\Delta T}^c$, $\mathbf{\Delta R}^c$, $\mathbf{\Delta t}^c$	Relative transformation, orientation, and displacement between cameras $\mathcal{C}^c$ and $\mathcal{C}^r$
$\mathbf{\Delta T}^{c\to d}$, $\mathbf{\Delta R}^{c\to d}$, $\mathbf{\Delta t}^{c\to d}$	Relative transformation, orientation, and displacement between cameras $\mathcal{C}^c$ and $\mathcal{C}^d$
$\mathbf{R}_{\mathbf{n},\mathbf{r}}$	Orthonormal matrix constructed from two vectors $\mathbf{n}$, and $\mathbf{r}$
$\mathbf{R}_{\mathbf{r},\theta}$	Rotation matrix with rotation axis direction $\mathbf{r}$ and angle θ
$\epsilon_{\parallel}, \epsilon_{\perp}$	Parallel and perpendicular part

Further Symbols

$\mathcal{G} = (\mathcal{V}, \mathcal{E})$	Undirected simple graph with vertices $\mathcal{V}$ and edges $\mathcal{E}$
$\mathbf{0}_{3\times1}$	Zero vector of dimension three
$\mathrm{diag}\,(\cdot)$	Diagonal matrix
$\mathbf{I}_{3\times3}$	Identity matrix of dimension three
$[\cdot]_\times$	Skew-symmetric matrix related to the cross product
$(\cdot)^T$	Transpose
$\lvert\cdot\rvert$	Determinant
$\lVert \cdot \rVert_2$	Euclidean norm

1 Introduction

Surround sensing is a key prerequisite for automated vehicles and mobile robots to operate in unconfined environments. Accurate information about the relevant static and dynamic environment is required at all times and in all situations in order to allow for safe operation. To achieve this goal, information from different, complementary sensors is usually combined [7, 28]. Advanced driver assistance systems for instance, which can be found in many automobiles today, typically combine a radar and a camera system [8, 89]. While radar sensors provide accurate range and relative velocity measurements, the camera system performs various detection and classification tasks. The role of vision systems becomes increasingly important as currently no other sensor offers the same versatility [77]. For example, by changing the lens we can trade angular resolution for field of view, which qualifies cameras as a short and mid-range sensor [92, 89]. Typically, cameras on mobile robots and cars have no moving parts, making them durable and inexpensive to manufacsture. However, the major advantage of cameras lies in the various kinds of information that captured images provide. Extracting the information is one of the major challenges on the way towards automated driving. Yet, since the early deployment of camera-based advanced driver assistance systems for lane departure warning, more functions have been introduced successively [37, 22], allowing to extract a richer set of information about the surrounding scene. To capture the complete environment of a vehicle, either a single omnidirectional camera, i.e. a camera with a 360° field of view in the horizontal plane, or a distributed multi-camera setup can be employed. While omnidirectional cameras seem appealing due to the single camera body and viewpoint it is often difficult or undesirable to mount them in a position with an unobstructed field of view (e.g. on top of a mobile robot) in practical applications. For this reason, multi-camera systems are usually preferred. Furthermore, the offsets between the cameras can be advantageous, for example to estimate the absolute scale of the velocity [41]. An omnidirectional panorama image can be constructed from just two cameras equipped with fisheye (ultra wide-angle) lenses. An example is shown in Figure 1.1. Despite the seamless appearance, objects on the two meter wide stripe in front and behind the vehicle do not appear in the image. For this reason, typically four cameras are employed in practice. To fuse the information from multiple cameras and to relate geometric quantities in the image and the world, the mapping from the 3D world into a 2D image has to be known. The mapping comprises information about the pose (orientation and displacement) of the camera as well as the projection from the camera coordinate frame into the image. Camera calibration is the process (and the result) of estimating the underlying model parameters. We refer to the parameters as extrinsic and intrinsic parameters, respectively[1]. Calibration is generally a complex and time consuming process. While intrinsic calibration can be performed for each camera individually prior to its deployment, extrinsic calibration requires the cameras to be mounted to the respective vehicle or robot. In order to reduce the effort of calibration and to compensate for external factors such as temperature variations and mechanical stress which may cause the extrinsic calibration to become inaccurate over time, self-calibration is highly desirable. Self-calibration is the process of inferring the model parameters directly from observations without the need for special calibration procedures or equipment.

In this thesis we build the theoretic foundation for extrinsic camera self-calibration and present and evaluate a Kalman filter-based approach which relies solely on image data. To this end, we identify and combine different cues that provide information about the calibration parameters. Motion-based calibration is carried out by estimating the frame-to-frame camera motion using corresponding features in successive images. Planar motions are common among mobile robots and road vehicles and represent a degenerate case for motion-based calibration. We overcome this problem by leveraging scene constraints. In particular, we make use of the ground plane as a natural reference object. A novel algorithm for ground plane estimation is presented that is robust with respect to sparse as well as structural outliers and can be integrated seamlessly into Kalman filters. Large baselines and strong geometric distortions hinder establishing feature correspondences between the images of cameras with overlapping fields of view to a degree where they are not used for calibration [82, 36]. We compensate these distortions using prior knowledge about the scene and camera configuration. As a result, low complexity feature detectors and matchers can be employed.

In contrast to existing approaches (e.g. [75]) we employ a single extended Kalman filter with low state vector dimensionality which reduces the overall complexity. We evaluate the approach qualitatively as well as quantitatively using real-world data.

[1] Photometric camera calibration is not considered here.

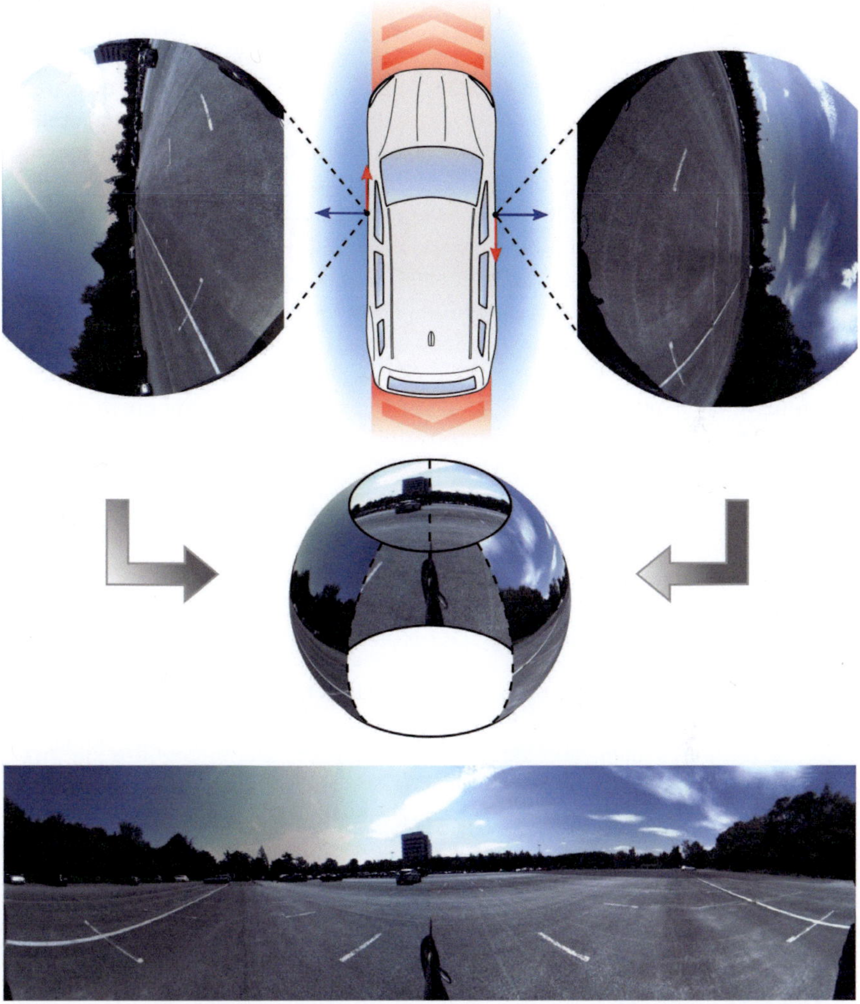

Figure 1.1: An omnidirectional panorama image is generated from images captured by a calibrated, vehicle-mounted two-camera setup. Despite the large camera offset of almost two meters, the panorama image appears seamless (except for image regions obstructed by parts of the vehicle). This is due to the the distant scene and homogeneity of the asphalt texture. The area marked in red in the top image is not visible in the image. The axes of the camera coordinate frames are colored red and blue, respectively.

1.1 Problem Statement

In this thesis we address the problem of continuous extrinsic self-calibration of vehicle-mounted multi-camera systems. Since any calibration will deteriorate over time, self-calibration is the only way to ensure reliable long-term operation. A self-calibration algorithm should be able to run in the background continuously and process incoming data as it arrives. It has to perform this task during regular operation and should hence leverage all available information sources. These include in particular motion, epipolar, and scene constraints.
Typical applications include experimental setups, end-of-line calibration, and recalibration during long-term operation. Therefore, an initial guess of the calibration parameters which can be obtained through simple external measurements should be sufficient for the algorithm to converge under normal circumstances. In addition, a versatile solution should work with different numbers of cameras and independently of other sensor modalities. Camera systems for surround sensing typically employ ultra-wide angle (e.g. fisheye) lenses. The calibration algorithm should thus be able to cope with strong geometric distortions.

1.2 Contribution

The contributions of this thesis are the following:

- We present a comprehensive analysis of several classes of motion, sensors, and algorithms with respect to degenerate cases. A combination of a class of motion, sensor, and algorithm is degenerated if the calibration yields ambiguous solutions. For these cases we present algorithms to determine the parameter values of the subset of unambiguous parameters. Additionally, we derive a criterion to identify degenerate camera configurations which cannot be calibrated using overlapping fields of view.
- To compensate for large viewpoint variations as well as geometrical distortions caused by fisheye lenses we introducc an image preprocessing step that uses prior knowledge about the camera configuration and scene geometry. Images are warped prior to extracting feature point correspondences in order to establish image similarity. In turn, low complexity feature detectors and matching algorithms can be employed.
- A novel ground plane estimation algorithm for fisheye cameras is presented which is designed to be robust with respect to sparse outliers among putative image correspondences as well as to structural outliers such as

other planes in the scene. The algorithm can be integrated seamlessly and efficiently into Kalman filters.

- A new algorithm for extrinsic camera self-calibration is presented and evaluated. The algorithm is based on Kalman filtering which provides flexibility with respect to additional information sources and renders real-time processing possible.
- In an extensive evaluation we assess the new extrinsic self-calibration algorithm quantitatively using real-world data. We compare different motion models, varying frame rates, and evaluate the influence of overlapping fields of view.

1.3 Thesis overview

The thesis is structured as follows. In **Chapter 2** we review related work. The chapter is partitioned with respect to the constraints that are imposed to estimate the calibration parameters. The constraints are fundamental to the calibration process and specific to the application, camera configuration, and environment. Numerous approaches, including the one presented in this thesis, impose multiple constraints. Here, we focus on the ones which most fundamental to each approach.

In **Chapter 3** we introduce the fundamentals of perspective (standard) cameras and highlight the differences to cameras equipped with fisheye lenses. We focus in particular on the geometric and photometric characteristics. The chapter closes with a brief introduction to two-view geometry.

Chapter 4 provides the theoretic foundation for extrinsic camera calibration. The methods and approaches to calibrate a multi-camera systems are diverse, but rely only on a small number of constraints. In this chapter, we introduce the concepts of motion-based calibration, the ground plane, and calibration using overlapping fields of view (Figure 1.2). The identification of degenerate cases is an important aspect of calibration. The detection of such cases is difficult in practice since measurement noise makes any system appear observable. A theoretical analysis of specific scenarios enables detecting degenerate cases prior to a practical or simulated evaluation.

The estimation of camera motion, the ground plane, and the relative pose between rigidly coupled cameras proposed in this thesis relies on image point correspondences. To compensate for the large extent of multi-camera setups and resulting viewpoint variations, as well as geometrical distortions caused by fisheye lenses, we propose an image warping step in **Chapter 5**. Captured images

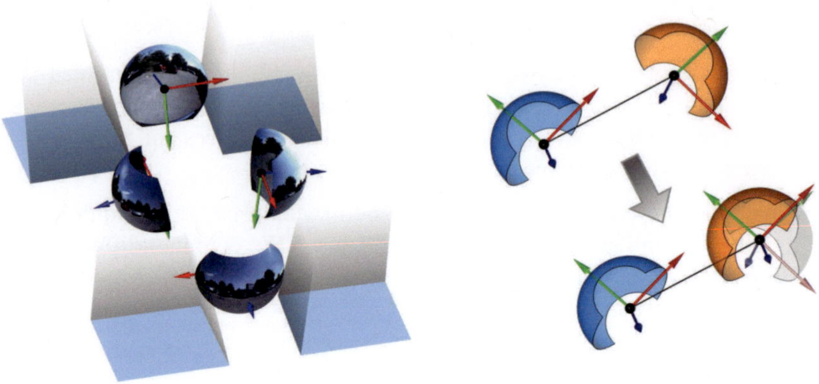

Figure 1.2: Left: Multi-camera setup with overlapping fields of view. To illustrate the fields of view of fisheye cameras image data is projected onto spherical sectors. The boundaries of the overlapping fields of view are indicated by blue patches on the ground plane and grey planes elsewhere. Right: Virtual camera rotation. To simplify establishing image point correspondences one of the cameras is virtually rotated. As a result, the position of infinitely distant objects coincides in both images. Throughout this thesis the reference cameras is colored orange, other cameras are colored blue.

are warped into virtual camera views such that corresponding image regions coincide. To this end, the scene geometry is approximated by the ground plane in close proximity and by infinitely distant objects elsewhere. In the latter case, the warped image corresponds to that of a virtually rotated camera (see Figure 1.2). As a result, low complexity feature detection and matching algorithms can be employed.

In **Chapter 6** we present an algorithm for robust ground plane estimation. The proposed method is designed to be robust with respect to sparse gross outliers but also to other structures in the scene with similar parameters. So called structural outliers such as sidewalks are hard to identify due to their inner coherence and may introduce significant bias. Given an estimate of the camera motion and ground plane we sample image point correspondences between successive images, starting with correspondences that exhibit the highest probability to be associated correctly and update the motion and ground plane estimate sequentially. The presented sequential testing and updating scheme is designed to be seamlessly integrable into Kalman filters. Figure 1.3 shows a comparison between a standard and the proposed approach.

Figure 1.3: Side-by-side comparison between the output of a standard approach for ground plane estimation (left) and the proposed method (right) for an image captured with a side-facing, vehicle-mounted fisheye camera. In the left image two planes are detected. Image point correspondences associated with either plane are marked orange and green. While one of the estimated planes can be associated with the sidewalk the other one does not correspond to any real plane in the scene. In the right image the output of our method is shown. Most of the shown image point correspondences (blue) are located on the road, as desired.

In **Chapter 7** we combine the findings and methods introduced in the previous chapters and present an algorithm for extrinsic camera self-calibration. The algorithm is based on an extended Kalman filter which has been applied successfully in similar scenarios. The state vector of the Kalman filter comprises the extrinsic calibration parameters as well as the ground plane parameters and the parameters of the applied motion model. A planar and a general one are considered. By applying a stratified update scheme, a partially updated state vector is made available for robust ground plane estimation.
In **Chapter 8** we evaluate the proposed extrinsic self-calibration algorithm using real-world data from a vehicle-mounted multi-camera system. The results are assessed quantitatively using acquired ground truth. Ground truth facilitates the comparison between the different motion models, algorithm settings, and information sources. During evaluation the algorithm is initialized using a set of 20 calibration parameter vectors that have been generated through random sampling (Figure 1.4a). Quantitative results are obtained for all permutations of the 20 initial parameter sets and 20 evaluation sequences. In addition, we present qualitative results using three typical applications for multi-camera systems, namely visual odometry, generation of a virtual top view of the vehicle surrounding, and stereo rectification. An example is shown in Figure 1.4b.

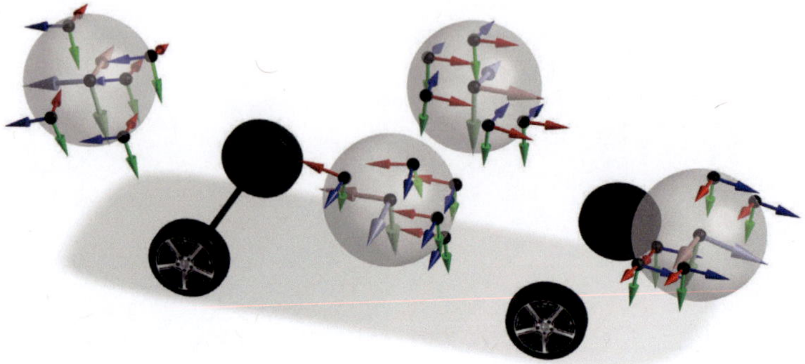

(a) Ground truth camera poses and samples for initialization.

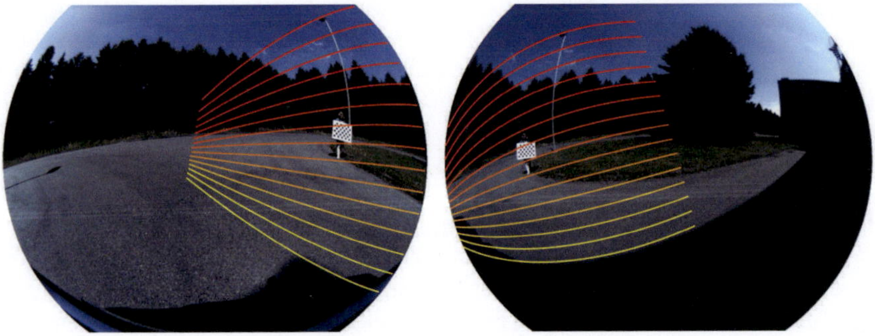

(b) Corresponding epipolar curves.

Figure 1.4: Ground truth camera poses and a subset of drawn samples for Kalman filter initialization (top). Large coordinate axes indicate the ground truth camera poses. Smaller coordinate axes visualize a subset of the initialization samples which are offset by 0.5 meters (transparent spheres) and rotated by up to 15° with respect to the ground truth. For reference vehicle tires and the rear axle are superimposed. In the bottom figure simultaneously captured images from the front (left) and right-facing (right) cameras are shown, respectively. Corresponding epipolar curves are superimposed. Matching curves have the same color.

Finally, in **Chapter 9** we summarize our work, highlight important findings, and discuss potential directions of future research.

2 Fundamentals

Multi-camera systems are employed in increasing numbers and more areas of everyday life. The methods and approaches used to calibrate these systems are as diverse as their respective fields of application. In this chapter we review relevant approaches from the literature and state-of-the-art solutions. We focus in particular on mobile robots and road vehicles.

We structured the related work presented in this chapter predominantly with respect to the underlying constraints that are imposed to estimate the calibration parameters. These constraints are fundamental to the calibration process and specific to the application, camera configuration, and environment (although other criteria for categorization could be applied as well). Figure 2.1 shows a taxonomy of extrinsic multi-camera calibration, without claiming thoroughness or completeness.

The fundamental assumption upon which all presented approaches build is the rigidity of the camera setup. The relative displacements and orientations between the cameras are assumed to be either fixed permanently, or within specific time frames[1]. From this, further constraints can be derived. Work focusing on simultaneously observed scene points is presented in Section 2.1, and work focusing on motion-based calibration and exploiting the scene structure is presented in Section 2.2 and Section 2.3, respectively.

Given an existing multi-camera setup, the constraints that can be applied are mostly predetermined by the physical arrangement of the cameras, the fields of view, the area of application, and the environment, leaving few design choices. However, one remaining aspect is the algorithm. In Section 2.4 we review the related work from the perspective of the underlying algorithm.

2.1 Overlapping Fields of View

The literature offers a plethora of works on the calibration of cameras in stereo configurations, i.e. with large overlapping fields of view. The standard approach

[1] Continuous parameter drifts are usually modeled by assuming changes to only occur between discrete points in time.

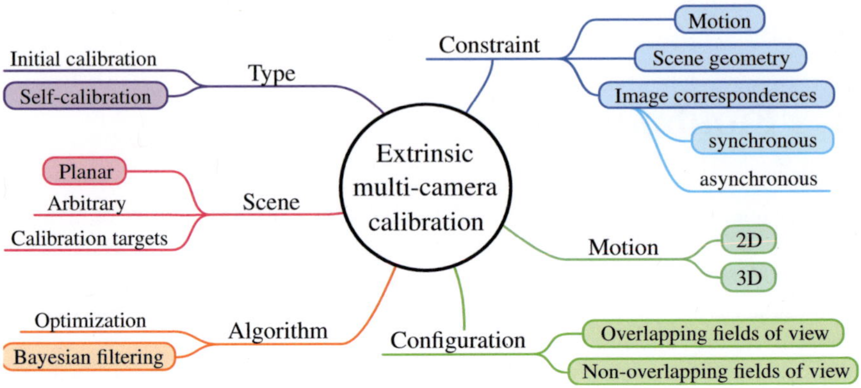

Figure 2.1: A taxonomy of extrinsic multi-camera calibration. The classification of the approach presented in this thesis is highlighted. The approaches often cannot be assigned uniquely to one class since multiple constraints or configurations might be exploited simultaneously.

to this problem is to establish (multi-camera) image correspondences. From these an initial estimate of the relative orientation and displacement can then be determined by means of relative pose estimators (e.g. [72]). Typically, the initial result is refined using bundle adjustment (e.g. [98, 31]). Bundle adjustment is a technique to simultaneously optimize the 3D scene structure and camera poses. The constraint underlying the relative pose estimation is the epipolar constraint (see Chapter 3) while bundle adjustment is based on the collinearity equations [57] which state that the camera projection center, image point, and 3D scene point were aligned at the time of recording.

A limitation of approaches employed in unconstrained environments is that the overall scale of the system cannot be determined without additional extraneous information and therefore remains ambiguous. To overcome this limitation and to simplify the process of establishing image correspondences customized calibration objects with known dimensions are commonly used. The description and discussion of algorithms working in unconstrained environments as well as with calibration objects can be found in standard literature such as [31] and are not discussed here for brevity.

Nonetheless, we want to highlight the work of Dang et al. [18] who presented a framework based on Kalman filtering for continuous stereo self-calibration of an active stereo system. It is shown that the combination of different constraints

(two-view epipolar and multi-view collinearity) yields both accurate and robust results. In the work presented herein we follow this idea and combine different constraints to improve the overall robustness and accuracy.
In general, calibration methods that exploit overlapping fields of view achieve the highest accuracy. To enable applying the same methods and constraints to multi-camera setups with non-overlapping fields of view, Kumar et al. [48] and Lébraly et al. [50] propose to create overlapping fields of view temporarily during calibration using mirrors. Kumar et al. [48] propose to install planar mirrors to allow all cameras to observe a single calibration object. By varying the position and orientation of the mirrors different virtual view points of the calibration object are generated. The extrinsic calibration parameters can be estimated uniquely from the virtual camera poses. Later, Lébraly et al. [50] present a modified approach that uses markers which are attached to the mirror to estimate the pose of the mirror directly.

Asynchronous Image Correspondences

From the perspective of calibration, image correspondences between synchronously captured images are most preferable since the underlying epipolar geometry comprises the desired calibration information. However, such correspondences can only be established between cameras with overlapping fields of view. Asynchronous image correspondences, on the other hand, can be established if the cameras observe the same part of the scene, but not necessarily at the same time. Prerequisites for the calibration based on asynchronous image correspondences are that the motion of the camera setup is known or can be estimated, and the scene remains static during data acquisition.
An early approach adopting this concept for road vehicles is the work of Lamprecht et al. [49]. To calibrate a multi-camera setup with non-overlapping fields of view, first, the 3D positions of traffic signs with respect to the vehicle are estimated. Once the traffic signs leave the field of view of the camera their position is predicted using known vehicle motion. As the traffic signs enter the field of view of another camera the relative orientation and displacement between the cameras is estimated by minimizing the error between the predicted and observed traffic sign positions.
The work of Carrera et al. [14] generalizes this concept. While a robot performs a set of preprogrammed motions the attached cameras separately estimate their motion and reconstruct the 3D scene. The reconstructions are then registered and jointly optimized providing the relative pose between the cameras. Due to the scale ambiguity of the monocular scene reconstruction, the displace-

ment between the cameras can be estimated only up to scale. Heng et al. [36] present a further extension to this approach. The trajectories estimated by each vehicle-mounted camera individually are registered with respect to the vehicle-supplied trajectory in order to obtain an initial calibration and scale estimate. Image point correspondences between asynchronously captured images are then used to refine the initial extrinsic calibration estimate along with the camera intrinsics using bundle adjustment. The scene, which was reconstructed during the calibration can itself be used as a calibration object. Heng et al. [34, 35] follow this idea to calibrate vehicle-mounted multi-camera setup with non-overlapping fields of view. Similarly, Li et al. [53] use a large calibration object that can be observed partly by multiple cameras at the same time. Strauss et al. [93] propose combining multiple planar calibration targets into a rigid, three dimensional calibration object. The calibration parameters and relative poses between the calibration objects are estimated jointly using bundle adjustment. Due to the known scale of the calibration objects these methods yield a Euclidean calibration[2].

2.2 Motion-Based Calibration

Motion-based extrinsic multi-camera calibration builds on the rigid coupling between the cameras and in particular on the different types of motions observed when the setup is moved. Due to the resemblance of the underlying mathematical formulation of the problem to a calibration problem in the robotic community between a robot gripper and a gripper mounted camera, this problem is often referred to as hand-eye calibration [100, 87]. An early work in the context of motion-based camera to camera calibration is that of Luong and Faugeras [58], who estimate the extrinsic calibration of a stereo camera without using overlapping fields of view. While the camera setup is moved, each camera estimates its motion. The extrinsic calibration between the cameras is then estimated up to scale from only two incremental motions by solving the hand-eye calibration problem explicitly.

Esquivel et al. [19] propose a similar approach but aim at processing complete sequences. In addition, critical motions such as translation only or planar motion are examined and the authors recommend switching the motion model if degenerated cases are detected.

[2] We use the term Euclidean, i.e. with known scale, to distinguish from metric calibration, i.e. with respect to a similarity transformation.

Muhle et al. [70] approach the problem of critical motions by incorporating a priori knowledge. The a priori knowledge ensures that the underlying optimization problem is well-conditioned. The authors further introduce a metric which quantifies the influence of the a priori knowledge on the final estimate and a transformation to remove the bias introduced by the a priori estimate.
Lébraly et al. [51] focus explicitly on planar motion and present a dedicated solution to the problem. Instead of using incremental motion estimates the camera motions and scene structure are estimated jointly. After determining an initial solution the relative orientation and in-plane translation are estimated using bundle adjustment.
We exemplarily mention the work of Brookshire and Teller [9, 10] who presented a modified solution to the hand-eye calibration problem for arbitrary sensors that provide Euclidean incremental motion estimates. To detect singular motions a statistical measure is used which provides a lower bound on the calibration accuracy. This concept is applied to both in-plane motions [9] and general motions [10]. In addition, they also propose a solution for sensor systems that provide data asynchronously.
Caspi and Irani [15] relax the requirement of known temporal alignment. By finding the maximum correlation between rotation amplitudes the temporal offset between two image sequences is found. After aligning the video sequences temporarily the relative orientation between the cameras is estimated.
Pagel et al. consider a similar setup to the one we examine herein. In a series of works [76, 74, 73, 75] they present a hierarchical approach based on repeated parameter estimation, propagation between camera modules, and fusion. After applying a method similar to that of Lébraly et al. [51] to obtain and register motion estimates a Kalman filter derivative is used to simultaneously refine extrinsic calibration parameters, sparse scene structure, camera motion, and ground plane estimates. The final estimate is obtained by fusing the individual estimates from each camera module. It is implicitly assumed that the camera translation directions are parallel and the relative camera velocity ratios remain constant within short time periods. In this thesis, we follow the idea of a filtering-based approach but reduce the algorithm complexity by employing only a single extended Kalman filter with low state vector dimensionality and relax the requirements on the vehicle motion.

2.3 Scene Constraints

The scenes in which multi-camera systems are deployed commonly contain cues that can be exploited for calibration. Road and parking spot markings have been used extensively hitherto (e.g. [54]). Many approaches, including the one presented herein, assume the surface the vehicle is driving on to be sufficiently flat to be approximated to be a plane in the vicinity of the vehicle. A single scene plane such as the ground plane constrains three out of the six degrees of freedom of the relative pose transformation (two angles and one distance), and does not necessitate overlapping fields of view.

Miksch et al. [66] propose to estimate the ground plane during straight driving by first estimating the vehicle translation direction. After an image rectification step that aligns image rows with the translation direction, only two corresponding points on the ground plane have to be identified in two successive images to find the ground plane parameters. The relative orientation between multiple cameras can be computed by aligning the observed ground plane normal, height, and translation directions.

Ruland et al. [82] estimate the in-plane position of a camera with respect to a vehicle frame by exploiting the non-holonomic motion of typical automobiles and estimating the ground plane induced homography.

The problem of estimating the orientation of camera with respect to a vehicle frame is closely related to homography estimation. For example, Miksch et al. [67] and Ruland et al. [81] present approaches in this regard using known vehicle odometry. An overview of several approaches, without focusing on calibration, is given in Chapter 6.

2.4 Bayesian Filtering and Optimization

In the remainder of this chapter we elaborate on the algorithms used to estimate the calibration parameters from image measurements and imposed constraints. We can classify these algorithms into general optimization and filtering techniques.

General optimization techniques such as bundle adjustment perform batch optimization using either all available measurements (global optimization) or specific subsets such as a fixed number of recent measurements (local optimization). In contrast, (Bayesian) filtering techniques fuse image measurements sequentially by updating the estimate and the associated probability distribution accordingly.

Bundle adjustment is considered the gold standard ([31]) and is known to better cope with nonlinearities and outliers than filtering-based approaches. For this reason, it is frequently used in offline calibration methods with mild resource and time constraints (e.g. [51, 14, 36, 93, 103]).
Self- and online-calibration problems are naturally incremental and are therefore traditionally approached using filtering techniques (e.g. [18, 30, 85, 75, 69]). However, the development of efficient and incrementally working optimization frameworks (e.g. [40, 46]) renders their application possible even for this type of application. The problem of increasing number of measurements can either be tackled by continuously summarizing measurements and results, as in filtering approaches, or by keeping only a subset of measurements and discard the remainder. The subset typically consists of a limited number of most recent observations. In contrast, Maye et al. [63] present a framework for self-supervised data aggregation that selects a subset of data based on an information theoretic measure. Despite the advantages of information-based measures these approaches tend to be particularly susceptible to outliers which spuriously indicate a high gain in information.
In this work we present a filtering-based approach that avoids structure computation entirely (except for the ground plane), and thus significantly reduces the overall complexity. The state vector of the employed extended Kalman filter comprises only the extrinsic calibration, ground plane, and motion parameters. Special attention is paid to the problem of outliers (Chapter 5 and Chapter 6) as well as the problem of nonlinearities (Chapter 7).

3 Camera Model and Two-View Geometry

A camera maps the 3D world into a 2D image. The mapping comprises information about the pose (orientation and displacement) of the camera coordinate frame with respect to a reference coordinate frame (e.g. a world frame) as well as the projection from the camera coordinate frame into the image. The goal of this thesis is to recover the former, the location and orientation of a camera coordinate frame with respect to a reference frame while considering the properties of fisheye cameras. Compared to cameras with standard lenses, the imaging properties of fisheye cameras differ in both their geometric and their photometric characteristics. This chapter gives an overview of these fundamental differences. First, a standard camera model is introduced in Section 3.1 which is then used to elaborate on fisheye cameras in Section 3.2. Parameters associated with the camera model are referred to as intrinsic calibration parameters. In contrast, the extrinsic calibration parameters describe the external geometric relation between the camera coordinate frame and the reference frame.
In the second part of this chapter, two fundamental (extrinsic) geometric relations of two-view geometry are reviewed, namely the plane induced homography and the essential matrix. Both will be used frequently throughout this thesis. In the following only rotationally symmetric camera models are considered. More literature on camera models can be found in, e.g., [31, 23, 94].

3.1 The Perspective Camera Model

The mapping from the camera coordinate frame into the image is described by the camera model. A model of particular interest is the perspective camera model[1] [31]. On the one hand, many real cameras can be described by this model directly or by adding correction terms. On the other hand, its mathematical formulation is particularly simple due to its linearity in homogeneous coordinates.

[1] Sometimes referred to as finite projective model.

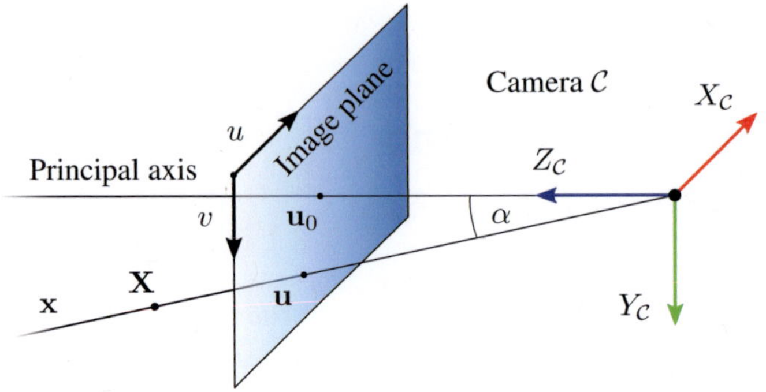

Figure 3.1: Projection of a 3D point $\mathbf{X}$ to the image point $\mathbf{u}$ under the central projection. The ray $\mathbf{x}$, originating from the optical center of camera $\mathcal{C}$ contains $\mathbf{u}$ and $\mathbf{X}$. The image point $\mathbf{u}$ and the principal point $\mathbf{u}_0$ lie in the image plane, depicted in blue. The origin of the image coordinate system is located in the top left corner and the image coordinate axes u and v are aligned with the axes of camera coordinate frame. The angle between the principal axis and $\mathbf{x}$ is the off-axis angle α. The axes of the camera coordinate frame are given in red, green, and blue, respectively. This color convention is kept throughout this thesis.

For this reason it is used extensively as a standard model in theoretical considerations. The camera model will be explained in more detail in the following. We consider the central projection, i.e. the projection from a point onto a plane, depicted in Figure 3.1. The projection center of a camera $\mathcal{C}$ coincides with the origin of the Cartesian camera coordinate frame. Within the image plane, $Z_\mathcal{C} = z_0$, where $z_0 > 0$, we define a 2D Cartesian image coordinate frame with coordinates u and v. The u and v-axes are parallel to the $X_\mathcal{C}$ and $Y_\mathcal{C}$-axes of the camera coordinate frame, respectively. The principal axis intersects the image plane in the principal point $\mathbf{u}_0 = (u_0, v_0)^T$. The three intrinsic calibration parameters u_0, v_0, and z_0 are sufficient to define the mapping of a 3D point $\mathbf{X} = (X, Y, Z)^T$ in the camera coordinate frame to the point $\mathbf{u} = (u, v)^T$ in the image.

Using homogeneous coordinates [31] the mapping can be written as a linear mapping

$$\begin{pmatrix} \lambda u \\ \lambda v \\ \lambda \end{pmatrix} = \underbrace{\begin{bmatrix} z_0 & 0 & u_0 \\ 0 & z_0 & v_0 \\ 0 & 0 & 1 \end{bmatrix}}_{=:\mathbf{C}} \begin{pmatrix} X \\ Y \\ Z \end{pmatrix}, \tag{3.1}$$

where $\mathbf{C}$ is the camera calibration matrix, and $\lambda \in \mathbb{R} \setminus \{0\}$ is a scale factor. The 3-vector $(\lambda u, \lambda v, \lambda)^T$ represents the point $\mathbf{u}$ in homogeneous coordinates. The non-homogeneous 2-vector can be obtained by dividing by λ and discarding the last row. Due to the homogeneous representation, equation (3.1) holds for any nonzero multiplicative scaling of $\mathbf{C}$ as well as $\mathbf{X}$. In Euclidean space, the scaling of $\mathbf{X}$ can be interpreted as the shift along the line through the origin and $\mathbf{X}$. Every point on the line, except for the origin, is projected to $\mathbf{u}$. This also means that a point with negative Z value is projected to real image coordinates. This does not generally pose a problem for perspective cameras since we know that a point being visible in the image has to be in front of the camera. However, if camera lenses exhibit an angle of view of more than $180°$, as it is the case for the cameras used during our experiments, disambiguating in this way is no longer possible. To resolve this issue we restrict the scale factors to positive values only. In consequence, only points on the *ray* $\mathbf{x} = (x, y, z)^T$, are projected to $\mathbf{u}$ (cf. Figure 3.1). The restriction of the sign of the scale factor to positive values allows discriminating between points on either side of the plane $Z_C = 0$, but requires to keep track of the sign during computation.

The presented ideal camera model is linear in homogeneous coordinates. However, real lenses and especially wide-angle lenses do not exhibit these linear characteristic. Typically, an image compression can be observed with increasing distance from the principal point for wide-angle lenses. The effect is most obvious for straight lines in the world appearing curved in the image. These radial distortions can be modeled by augmenting the linear model by a correcting term. The correction is applied after projecting the 3D point into the image. Let $\mathbf{u}_u = (u_u, v_u)^T$ be the coordinates of the undistorted point in the image plane. The coordinates of the point in the radially distorted image are given by

$$\mathbf{u}_d = \mathbf{u}_u \left(1 + \Delta\left(r\right)\right), \tag{3.2}$$

where $\Delta(\cdot)$ is typically chosen to be an even polynomial [31, 38, 23], and $r = \|\mathbf{u}_u - \mathbf{u}_0\|_2$ is the radius, i.e. the distance from the principal point.

Correction terms $\Delta(\cdot)$ are used multiple times throughout this chapter to model deviations from a design model.

3.2 The Fisheye Lens

Fisheye lenses are ultra wide-angle lenses that are often capable of capturing a whole hemisphere. The large field of view comes at the price that many desirable properties of the perspective camera cannot be obtained, most prominently linearity is lost. Straight lines in the world are not imaged as straight lines.
This section gives an overview of fisheye lens characteristics, starting with the geometric camera model.

3.2.1 Geometric Camera Model

Due to the large field of view it is not possible to model the nonlinear properties of cameras with fisheye lenses as deviations from the linear model, i.e. lens distortions. To demonstrate this, we consider the projection of a point $\mathbf{X} = (X, Y, 0)^T$ into the image using equation (3.1). After multiplying the point with the calibration matrix, the last component of the point remains zero. It is not possible to convert the point to finite coordinates and thus to Euclidean 2-space. In consequence, equation (3.2) cannot be applied. For fisheye lenses it is common to describe the mapping from the world into the image in terms of spherical coordinates, i.e. by the off-axis angle α between the ray $\mathbf{x}$ and the principal axis (cf. Figure 3.1), and the azimuth angle in the image plane.
In case of rotational symmetry, there is a direct relationship between the radius r and the off-axis angle α. For example, the standard camera model follows $r \sim \tan(\alpha)$. For fisheye lenses there exist various classical design models that exhibit specific properties [94]. Three prominent models are the

- stereographic model $\quad r \sim \tan\left(\frac{\alpha}{2}\right)$,
- equidistant model $\quad r \sim \alpha$,
- and equisolid angle model $\quad r \sim \sin\left(\frac{\alpha}{2}\right)$.

The functions are shown in Figure 3.2. The stereographic mapping is locally distortion free, i.e. within a sufficiently small region objects are imaged as being captured by a perspective camera with a narrow field of view. Furthermore, the intersection angle of imaged lines is only affected by perspective distortions, but not by lens distortions. Hence, the mapping preserves angles locally.

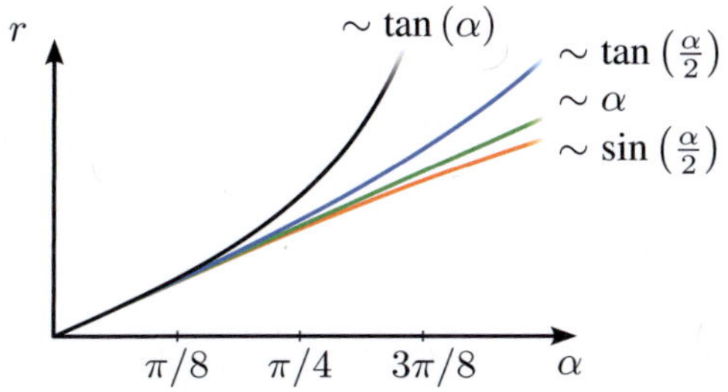

Figure 3.2: Radius r as a function of the off-axis angle α for four different projection models. The projection models are (from top to bottom), perspective model (black), stereographic model (blue), equidistant model (green), and equisolid angle model (orange). In case of the perspective mode, the radius approaches infinity as the off-axis angle approaches $\pi/2$.

The equidistant mapping function is linear in the off-axis angle, and the equisolid angle model maintains a constant ratio between image area and corresponding solid angle. Figure 3.3 shows an image captured with a fisheye lens with equidistant projection function and a corresponding image which has been generated using a perspective camera model. Note that the white marking on the ground is straight in the perspective camera image but not in the image of a fisheye camera. Furthermore, significant magnification near the image boundary can be observed.

Mapping functions of real fisheye lenses may deviate from the ideal models. To compensate for this behavior polynomial correction terms as in equation (3.2) can be applied. Throughout the rest of the thesis, we abstract from the used model and intrinsic calibration parameters and write

$$\mathbf{u} = \boldsymbol{\kappa}\left(\mathbf{X}\right) = \boldsymbol{\kappa}\left(\mathbf{x}\right), \tag{3.3}$$

to describe the projection of a 3D point or corresponding ray into the image, where $\boldsymbol{\kappa}\left(\cdot\right)$ is the projection function. The back-projection of a point to a ray is given by

$$\mathbf{x} = \boldsymbol{\kappa}^{-1}\left(\mathbf{u}\right). \tag{3.4}$$

To render the back-projection to a ray unique, one typically requires $z = 1$ or $\left\|\mathbf{x}\right\|_2 = 1$. We further require the mapping to be well-defined within the image region and to be continuously differentiable.

Figure 3.3: Image captured using a fisheye lens with equidistant projection (left), and corresponding image generated using a perspective camera model. The projection of the image border of the perspective image mapped into the fisheye image is shown in orange. The horizontal angle of view is 170° and 100°, respectively.

3.2.2 Noncentrality

The fundamental assumption for the derivation of the ideal perspective camera model in Section 3.1 was the existence of a unique projection center. This property is highly desirable, as it allows separating intrinsic and extrinsic camera properties. However, for real lenses the position of the projection center may deviate with increasing off-axis angle. For lenses with a narrow field of view, the effect is usually small and thus often disregarded. However, for fisheye lenses, the deviation can be within the same order of the size as the lens [23]. In the following we introduce a mathematical model for the deviation of the projection center and show how the noncentral camera model can be approximated by a central camera model with minimal error.

For rotationally symmetric lenses, the deviation of the projection center can be modeled by a displacement along the principal axis [23]. An illustration is shown on the left-hand side of Figure 3.4 for rays with increasing off-axis angles. The point $\mathbf{c}_0$ is the convergence point for decreasing off-axis angles. For increasing off-axis angles, the projection center moves forward along the principal axis.

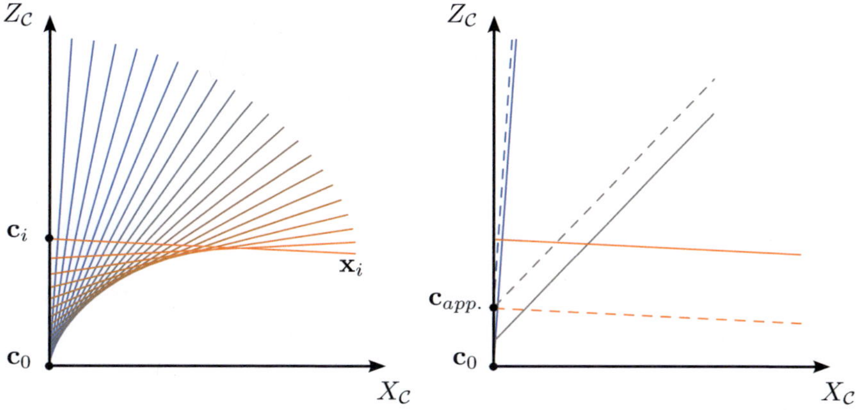

Figure 3.4: Illustrations of the displacement of the projection center along the principal axis in the X_C-Z_C-plane of the camera coordinate frame for rays with small (blue) and large (orange) off-axis angles (solid lines). The point $\mathbf{c}_0$ is the convergence point for decreasing off-axis angles. Left: A ray $\mathbf{x}_i$ originates from a displaced projection center $\mathbf{c}_i$. Right: The noncentral camera model is approximated by a central camera with projection center $\mathbf{c}_{app.}$. Corresponding rays originating from the $\mathbf{c}_{app.}$ are shown as dashed lines. Note that corresponding rays are parallel.

Gennery [23] proposes modeling the displacement as $\mathbf{c}(\alpha) = (0, 0, c_z(\alpha))^T$, where

$$c_z(\alpha) = \left(\frac{\alpha}{\sin(\alpha)} - 1\right)(\Delta_0 + \Delta(\alpha)), \tag{3.5}$$

$\Delta(\cdot)$ is an even polynomial, and Δ_0 is a constant. The first factor ensures that the displacement vanishes for small angles but increases to infinity as the off-axis angle approaches $180°$. The second factor determines the displacement magnitude. If the distance to the observed scene is sufficiently large, the error introduced by disregarding the deviation of the projection center becomes insignificant. For the calibration of large field of view cameras, however, it is common to use small calibration targets which are placed closely to the camera to achieve considerable coverage in the image (e.g. [64]). For this reason, using a central camera model during calibration will result in an incorrect mapping between the radius r in the image and the off-axis angle. This can be avoided by using a noncentral camera model during calibration and approximating the result by a central camera model. To this end, Schönbein et al. [86] propose to choose the projection center of the approximated central camera such that it minimizes the Euclidean distance to the rays corresponding to the image points of a uniformly sampled image. The process is illustrated on the right-hand side of Figure 3.4. In contrast to using a central camera model during calibration, the angular error of the approximated model decreases the farther an object is located from the camera. Throughout the rest of the thesis, we assume the camera model to be central and apply equations (3.3) and (3.4).

3.2.3 Light Falloff and Vignetting

Examining the image on the left-hand side of Figure 3.3 we notice a gradual reduction in image intensity towards the image boundary as well as a sudden transition to black in the image corners. These effects are caused by light falloff and vignetting and can be partially compensated. Light falloff is inherent to optical systems and is, for this reason, also referred to as natural vignetting. It is caused by light entering and exiting the optical system at oblique angles, less light entering the optical system, as well as the light being distributed over a larger area on the sensor. Under certain assumptions, the falloff is proportional to $\cos^4(\alpha)$ [38, 57]. In practice however, the assumptions were found to rarely apply, even for standard lenses, and in particular for fisheye lenses [3, 38]. The characteristic of the light falloff depends strongly on the lens design and should be determined though calibration.
Vignetting refers to the physical obstruction (which is not caused by the aperture stop). One commonly differentiates between three classes of vignetting [24]:

- Mechanical vignetting is caused by obstructing elements blocking incidental light before it can enter the lens. In the image on the left-hand side of Figure 3.3 the lens mount limits the diagonal angle of view and causes a sudden transition to black in the image corners.
- Optical vignetting refers to light being blocked by elements within the lens body such as edges or mechanical stops. Despite the downside, optical vignetting can also be used to improve the overall lens performance, e.g. by blocking misguided rays.
- Pixel vignetting is not caused by the lens but by the image sensor. Only a part of the total area of a pixel on the sensor is light sensitive. At oblique angles, it is more likely that light is blocked by obstructing elements on the sensor, yielding an angle dependent characteristic. To compensate for the insensitive area, micro lenses are commonly used to direct the incident light onto light sensitive area, thus effectively enlarging it. However,

Figure 3.5: Side-by-side comparison between the original image (left) before and after applying vignetting and light falloff compensation (right). Note the severe intensity reduction towards the image boundary in the original image, and the almost even intensity across the ground in the compensated image. In the compensated image reduced brightness around the clouds in sky can be observed. This is due to overexposure in the original image.

micro lenses may even reduce the angle range at which light is accepted [38]. Pixel vignetting is particularly prominent for wide-angle lenses [57]. This effect is increased if non-matching lenses and image sensors are combined.

The intensity reduction seen in Figure 3.3 is caused by a superposition of vignetting and light falloff. For image processing, the gradual reduction in image intensity can be disadvantageous. This is for example the case when the intensity gradient caused by light falloff and vignetting within a typically sized image patch becomes significant, or when comparing image patches from two images with opposing gradients. In general, it is possible to compensate for vignetting and light falloff due to the linearity of the effects [38]. Assuming an image sensor with linear response, i.e. a proportional relationship between irradiance and image intensity, the compensation can be carried out by pixel wise multiplication with a compensation factor. The compensation factor can be determined experimentally by measuring the pixel intensity in the image with respect to a constant illumination source. An exemplary result is shown in Figure 3.5. After compensation we observe an almost even intensity profile on the ground. Mechanical vignetting, however, cannot be compensated as no image information is available.

3.3 Two-View Geometry

In the remainder of this chapter we consider the extrinsic relations between two camera views. The fundamental relation between two perspective views of a scene is the epipolar geometry. It is independent of the scene content and depends only on the relative camera orientations and displacements as well as the camera intrinsic calibration parameters. It can be described concisely by the essential matrix. A second relation arises for 3D points being located on a plane in the scene. The plane induces a homography between perspective views, a one-to-one relation between image points. For the two relations to be meaningful, we assume the two views to be either acquired simultaneously or restrict the scene to be rigid in case that the views are acquired with a temporal offset. Both scenarios are geometrically equivalent [31]. Before elaborating on the two geometric relations we introduce the transformations between coordinate frames in 3-space.

3.3.1 Camera Pose and Pose Transformation

The pose of a camera encompasses the orientation and displacement of the camera coordinate frame with respect to a reference coordinate frame. Given the coordinates of a 3D point $\mathbf{X}$ in the camera coordinate frame, the coordinates of the same point in the reference coordinate frame $\mathbf{X}' = (X', Y', Z')^T$ are given by

$$\mathbf{X}' = \boldsymbol{\Delta}\mathbf{R}\mathbf{X} + \boldsymbol{\Delta}\mathbf{t}, \tag{3.6}$$

where $\boldsymbol{\Delta}\mathbf{R}$ is the 3×3 orientation matrix and $\boldsymbol{\Delta}\mathbf{t}$ is the 3×1 displacement vector. Using homogeneous coordinates, the transformation can be written more concisely using matrix notation

$$\begin{pmatrix} X' \\ Y' \\ Z' \\ 1 \end{pmatrix} = \underbrace{\begin{bmatrix} \boldsymbol{\Delta}\mathbf{R} & \boldsymbol{\Delta}\mathbf{t} \\ \mathbf{0}_{3\times 1}^T & 1 \end{bmatrix}}_{=:\boldsymbol{\Delta}\mathbf{T}} \begin{pmatrix} X \\ Y \\ Z \\ 1 \end{pmatrix}. \tag{3.7}$$

The transformations between coordinate frames can either be constant over time or time dependent. To emphasize the difference we use different notation and expressions. The time independent pose transformation between coordinate frames is given by equations (3.6) and (3.7) and will be called orientation and displacement. Similarly, the transformation between a coordinate frame at time k and $k+1$ is given by $\mathbf{T}_k$ and will be called rotation and translation. The rotation matrix and translation vector are $\mathbf{R}_k$ and $\mathbf{t}_k$, respectively. When necessary, a camera index c is used to differentiate between multiple cameras.

3.3.2 Epipolar Geometry and the Essential Matrix

Epipolar geometry is the inherent relation of two views. It is determined by the relative pose of the cameras and their intrinsic calibration parameters only, and independent of the scene. For a perspective camera the relation is encapsulated in concise form in the fundamental matrix. The fundamental matrix describes the relationship between an image point in one view and a corresponding epipolar line in the other view. However, for fisheye cameras, or other cameras with nonlinear projection function, the fundamental matrix cannot be applied and the relation becomes more complicated, typically resulting in a point to curve relationship in the image. A specialization of the fundamental matrix that separates intrinsic and extrinsic calibration parameters is the essential matrix. It formulates the two-view relationship for rays instead of image points and is thus applicable to arbitrary cameras with known intrinsic parameters. In the following we elaborate on epipolar geometry and derive the essential matrix.
Suppose a moving camera C that acquires image at time k and $k+1$. The two camera centers at time k and $k+1$ and a 3D point $\mathbf{X}$ define a plane which is called the epipolar plane. The rays back-projected from the image points of $\mathbf{X}$ are $\mathbf{x}$ and $\mathbf{x}'$, respectively. This is depicted in Figure 3.6. Using the rays and the transformation between the camera poses the epipolar plane in the second view can be constructed by computing the cross product between the ray $\mathbf{x}'$ and the translation vector

$$\mathbf{l}' = \mathbf{x}' \times \mathbf{t}_k. \tag{3.8}$$

The epipolar plane is then given by $((\mathbf{l}')^T, 0)^T$, where $\mathbf{l}'$ corresponds to the (non-unit) plane normal. Note that $\mathbf{l}'$ represents a line in 2D projective geometry and thus a homogeneous vector. The inner product of a point with $\mathbf{l}'$ is zero if the point lies on the plane, and by definition

$$0 = (\mathbf{l}')^T \mathbf{R}_k \mathbf{x}. \tag{3.9}$$

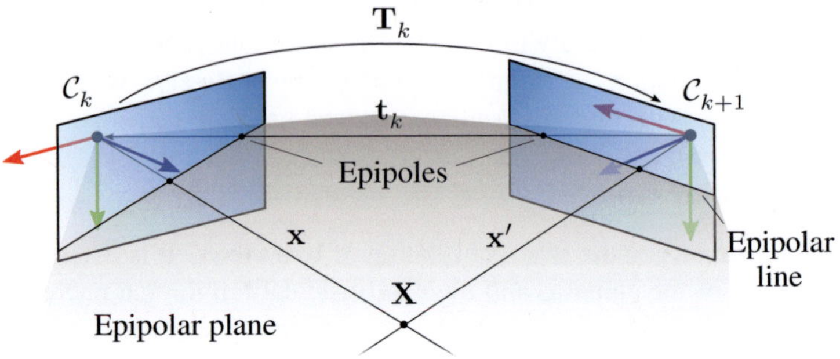

Figure 3.6: Epipolar geometry of a moving perspective camera. The epipolar plane contains the two camera centers as well as the 3D point $\mathbf{X}$. The intersections of the epipolar plane and the image planes form the epipolar lines.

Replacing $\mathbf{l}'$ in equation (3.9) by equation (3.8) yields

$$0 = (\mathbf{x}' \times \mathbf{t}_k)^T \mathbf{R}_k \mathbf{x} = (\mathbf{x}')^T \underbrace{[\mathbf{t}_k]_\times \mathbf{R}_k}_{=:\mathbf{E}_k} \mathbf{x}, \tag{3.10}$$

where $\mathbf{E}_k$ is the essential matrix, and $[\cdot]_\times$ is a mapping of a 3-vector to a skew-symmetric matrix

$$[\mathbf{x}]_\times = \begin{bmatrix} 0 & -z & y \\ z & 0 & -x \\ -y & x & 0 \end{bmatrix}. \tag{3.11}$$

The essential matrix is of rank two and has five degrees of freedom, namely the three parameters describing the rotation and two parameters describing the direction of the translation. Note that equation (3.10) holds for any nonzero scaling of $\mathbf{E}_k$ and hence of $\mathbf{t}_k$. The essential matrix can be estimated from five corresponding rays [72], however, yielding multiple solutions. For disambiguation additional correspondences are required. Furthermore, the decomposition of the essential matrix into a rotation and translation direction is also ambiguous [31]. However, throughout the rest of the thesis, we assume the correct decomposition to be known.

Equation (3.10) imposes only a single constraint on the rays $\mathbf{x}$ and $\mathbf{x}'$, i.e. the epipolar constraint. In the image of a perspective camera, the epipolar plane is imaged as a line called the epipolar line, intersecting the image of $\mathbf{X}$ and the camera center (cf. Figure 3.6). The image of the camera center is the called the epipole. In cameras with fisheye lenses the epipolar lines appear in general as curves.

3.3.3 Plane Induced Homography

If points in the scene are located on a plane, the corresponding rays of two views are related by a homography, $\mathbf{x}' = \mathbf{H}\mathbf{x}$. The homography matrix $\mathbf{H}$ is a non-singular 3×3 matrix and comprises information about the relative camera poses and the scene plane. It can be interpreted as the projection of a point onto the plane followed by the projection into the second view. In the following we derive the homography matrix.
A plane in Euclidean 3-space can be defined by the unit normal vector $\mathbf{n}$ and the distance to the plane h. Without loss of generality, we define $h \geq 0$. A 3D point $\mathbf{X}$ located on the plane satisfies

$$\mathbf{n}^{\mathrm{T}}\mathbf{X} + h = 0. \tag{3.12}$$

Given the ray $\mathbf{x}$ corresponding to the point $\mathbf{X}$ located the plane the correct scale λ can be found by substituting $\lambda\mathbf{x}$ for $\mathbf{X}$ in equation (3.12) and solving for λ,

$$\lambda = -\frac{h}{\mathbf{n}^{\mathrm{T}}\mathbf{x}}. \tag{3.13}$$

After determining the scale, the 3D point can be transferred from one coordinate frame to the other by applying the relative pose transformation (equation (3.6)). We assume, again, a moving camera C that acquires images at time k and $k+1$ (see Figure 3.7). By dividing both sides of the relative pose transformation $\mathbf{X}' = \mathbf{R}_k\mathbf{X} + \mathbf{t}_k$ by a scale factor and substituting for λ on the right-hand side, we obtain

$$\mathbf{x}' = \mathbf{R}_k\mathbf{x} - \frac{\mathbf{t}_k\left(\mathbf{n}_k\right)^T\mathbf{x}}{h_k} = \underbrace{\left(\mathbf{R}_k - \frac{\mathbf{t}_k\left(\mathbf{n}_k\right)^T}{h_k}\right)}_{=:\mathbf{H}_k}\mathbf{x}. \tag{3.14}$$

The homography matrix $\mathbf{H}$ has eight degrees of freedom and can be estimated from four corresponding image points [31]. The decomposition of the homography matrix into the rotation, translation and plane is ambiguous, yielding four possible solutions [59]. Besides two different solutions for the rotation matrix, the translation and the plane, one ambiguity is caused by the simultaneous change in the signs of $\mathbf{t}_k$ and $\mathbf{n}_k$ yielding the same homography matrix. Fur-

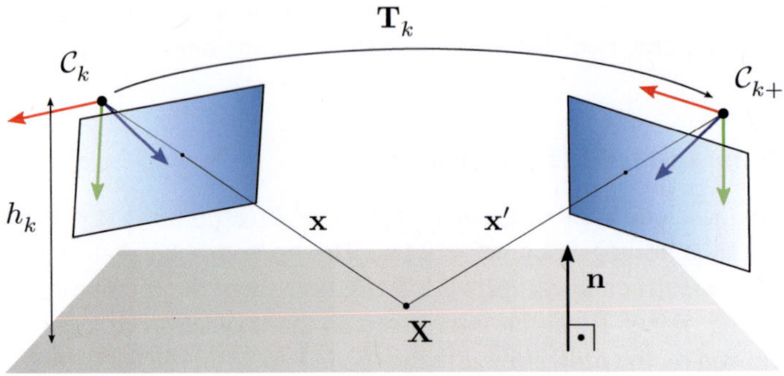

Figure 3.7: A moving camera acquires images at time k and $k+1$. The rays $\mathbf{x}$ and $\mathbf{x}'$, corresponding to the 3D point $\mathbf{X}$ on the scene plane, are related by a homography.

thermore, only the ratios $\mathbf{t}_k/h_k$ can be determined. The ambiguity is caused by a faster moving camera and a more distant plane resulting in the same homography as a slower moving camera and a closer plane. In the following chapters we assume the correct decomposition to be known in the following.

4 Extrinsic Camera Calibration

To estimate the extrinsic camera calibration several constraints can be used. Common are the epipolar constraint for simultaneously observed 3D points (cf. Chapter 3.3.2) and the rigid coupling between multiple cameras mounted on a rig. For self-calibration we combine several constraints to increase the robustness and to avoid degenerate cases. Pure translation, for example, renders motion-based calibration degenerate. Detecting degenerate cases is difficult in practice as measurement noise and errors in feature matching render classical tests such as rank analysis ineffective. Statistical measures [9, 10, 62] provide a means to detect degenerate cases if the underlying statistical assumptions hold. A theoretical analysis of specific scenarios based on error-free data allows detecting degenerate cases prior to a practical or simulated evaluation.
In this chapter, we present a comprehensive analysis of several classes of motion, sensors, and algorithms for motion estimation with respect to degenerate cases. The problem of detecting such cases is closely related to observability analysis in control theory. A system is called observable if its state can be recovered uniquely in finite time from its outputs and known inputs [6]. In the following we (informally) adopt the term to denote parameters whose values can be inferred. Our contribution with respect to motion-based calibration is twofold. We identify degenerate cases among the combinations of classes of motion, sensors and employed algorithms and, in addition, determine the observable parameters for degenerate configurations. Besides the rigidity constraint between cameras we incorporate the ground plane as a natural reference object into the analysis. As input we assume error free observations of the motion and ground plane parameters. The results are summarized concisely in Table 4.1.
In addition, we consider the extrinsic calibration of a multi-camera system from pairwise overlapping fields of view. Jointly observed 3D points allow estimating the essential matrix, and hence to recover the relative orientation and displacement direction. For a multi-camera setup a unique solution (up to an unknown scale factor) can only be determined if enough overlapping fields of view between different cameras exist and if the cameras are not in a critical configuration. To detect whether a unique solution (up to scale) can be derived, we employ a matrix rank test. Before presenting our analysis on motion-based ex-

trinsic multi-camera calibration we introduce necessary definitions. Parts of the work presented in this chapter have been published in [43].

4.1 Definition of the Reference Frame

The goal of this thesis is metric calibration of a multi-camera system. The term metric denotes that the calibration is unique up to a similarity transformation, i.e. a pose transformation and a scale. For the calibration process, the parameters corresponding to the seven degrees of freedom of a similarity transformation have to be defined by means of a datum definition[1]. The datum definition enables the mapping of relative observations onto absolute parameter values and is required to avoid singularities. The datum definition corresponds to defining an Euclidean (reference) coordinate frame as well as a scale. This can be carried out by defining one camera coordinate frame in a multi-camera setup as the reference coordinate frame (located at $\mathbf{0}_{3\times 1}$ and with identity orientation matrix) and keeping one baseline, i.e. the distance between two cameras, fixed. Minimal datum definitions, as in this example, which constrain exactly seven degrees of freedom are favorable as they avoid possible inconsistencies in the datum definition which could be misinterpreted as errors in the observations [57].
A disadvantage of the fixed datum is that the covariance matrix associated with the estimated calibration parameters of each camera does not reflect the inner accuracy of the camera system, i.e. the accuracy independent of the choice of reference coordinate frame [26, 97]. The position and orientation of the reference camera coordinate frame are assumed to be error-free, whereas other camera coordinate frames are subject to inaccuracies. For this reason, other datum definitions such as the free net adjustment are favored [57, 98]. In free net adjustment seven linear independent constraints are introduced that prevent perturbations of the centroid of camera centers, orientation, and scale with respect to initial (provided) estimates. However, gradual drifts caused by, e.g., numerical inaccuracies are not corrected. Typically, Lagrangian multipliers are used as a means to impose these constraints. The free net adjustment provides optimal inner accuracy [57].
It is possible to switch between different datum definitions without introducing errors if the corresponding parameter transformations are linear [98]. For this reason, Triggs et al. [98] propose applying a simple and convenient datum definition during estimation and apply an optimal datum definition afterwards,

[1] In the literature often the term gauge fixing is used instead of the geodesic term datum definition.

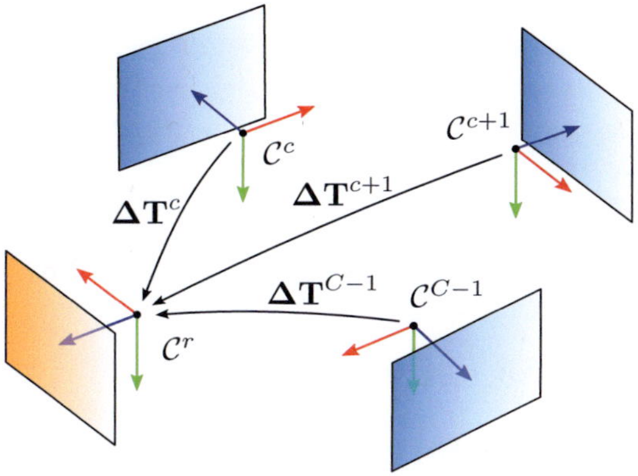

Figure 4.1: Schematic representation of the multi-camera system. Several camera coordinate frames are related to the reference coordinate frame via relative pose transformations. The image plane of the reference camera is shown in orange. This color convention is kept throughout the thesis.

thus, reducing the number of parameters and the computational cost. We adopt this approach and apply the minimal datum definition as presented in the example. To this end, a dedicated reference camera coordinate frame, denoted $\mathcal{C}^r$, is selected. All cameras $c = 0, ..., C-1$ are related to the reference camera coordinate frame via relative pose transformations $\Delta\mathbf{T}^c$ (cf. Section 3.3.1). This means in particular that a point $\mathbf{X}^c$ given in the coordinate system of camera $\mathcal{C}^c$ and a point $\mathbf{X}^r$ given in the coordinate system of the reference camera are related by

$$\begin{pmatrix} \mathbf{X}^r \\ 1 \end{pmatrix} = \Delta\mathbf{T}^c \begin{pmatrix} \mathbf{X}^c \\ 1 \end{pmatrix} . \tag{4.1}$$

The relation between the cameras is depicted in Figure 4.1. Without loss of generality, we define $r = 0$, such that $\Delta\mathbf{T}^r = \mathbf{I}_{4\times4}$. Furthermore, the reference coordinate frame is associated with the current pose of the reference camera, thus moving with the vehicle. The transformation between two cameras can be computed by concatenating and inverting pose transformations. For example, the transformation from camera $\mathcal{C}^1$ to $\mathcal{C}^2$ is given by $(\Delta\mathbf{T}^2)^{-1}\Delta\mathbf{T}^1$.
Different definitions of the scale are used in this thesis to allow for simple derivations. For example, for a vehicle moving in the plane parallel to the ground

plane the distance of the reference camera center to the ground is used to define the scale. Alternatively, the traveled distance of a camera center between two time steps could also be used. Some sensors, such as calibrated stereo cameras already provide a scale. In this case only the six degrees of freedom of the reference coordinate frame have to be constrained.
It should be noted that a comparison between calibration results is only possible if the same datum definition is applied. Gradual drifts, as in the case of the free net adjustment, have to be compensated [98].

4.2 Motion-based Calibration

In the following we examine the estimation of the extrinsic calibration parameters on the basis of the rigidity constraint between cameras for different classes of motion and, optionally, the ground plane. Four different classes of motion are analyzed, namely linear motion, circular motion, planar motion, and general (unconstrained) motion. The classes resemble the typical driving maneuvers straight driving, turning, and driving on planar ground. General motion takes further effects such as pitching, rolling, as well as nonplanar translation into account. For each class of motion, we present an algorithm to compute the observable parameters. This is done for both a multi-camera system of monocular cameras without overlapping fields of view, as well as a system of multiple cameras that provide instantaneous depth measurements (e.g. stereo cameras).
As input we assume error free observations of the rotations and translations, and optionally the ground plane normal and distance of the ground plane to the camera center (camera height). Monocular camera systems suffer from the problem of scale ambiguity, i.e. the scale of the translation cannot be recovered. For this reason we further distinguish between pairwise evaluation of consecutive frames, in which case we use the ground plane as a reference object, and using image triplets. By using the ground plane, the translation velocity with respect to the camera height can be recovered, which allows propagating information about relative velocities. This concept will be explained in more detail in this chapter. Image triplets allow propagating scale information by means of triangulating and reprojecting 3D points as in classical structure from motion approaches [31].

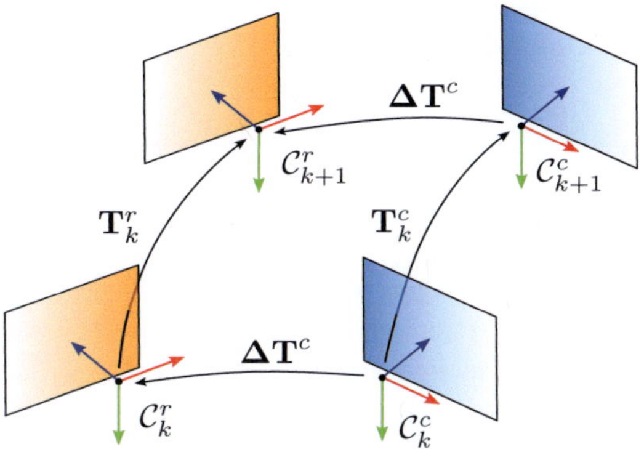

Figure 4.2: Hand-eye calibration. The transformations between the poses of a moving system of two rigidly coupled cameras form the characteristic circle of spatial and temporal relations.

4.2.1 Hand-Eye Calibration

Originally, hand-eye calibration referred to the estimation of the rigid relative pose between the coordinate frame of a camera mounted on the gripper of a robot and the coordinate frame of the gripper itself [100, 87]. To estimate the transformation, the gripper performs a known motion while the camera captures a known calibration object. The gripper motion, camera motion, and the unknown pose transformation form a circle of temporal and spatial transformations. Applied to a setup of two rigidly coupled cameras (instead of one camera and the gripper) we can write the concatenation of transformations in the characteristic form

$$\mathbf{T}^r_k \mathbf{\Delta T}^c = \mathbf{\Delta T}^c \mathbf{T}^c_k. \tag{4.2}$$

The circle of transformations is depicted in Figure 4.2. Due to the rigid coupling of the cameras $\mathbf{\Delta T}^c$ is constant over time. Equation (4.2) plays a fundamental role in motion-based extrinsic calibration and will be used extensively throughout this thesis. In contrast to the original problem, the motions of the cameras are not known and have to be estimated from observations.

Equation (4.2) can be decomposed into one equation relating rotations and orientations

$$\mathbf{R}_k^r \mathbf{\Delta R}^c = \mathbf{\Delta R}^c \mathbf{R}_k^c, \tag{4.3}$$

and one equation relating the displacement and translation vectors

$$\left(\mathbf{R}_k^r - \mathbf{I}_{3\times 3}\right) \mathbf{\Delta t}^c + \mathbf{t}_k^r - \mathbf{\Delta R}^c \mathbf{t}_k^c = \mathbf{0}. \tag{4.4}$$

In the following we discuss some properties of equations (4.3) and (4.4) in the context of motion-based camera calibration. For pure translational motion ($\mathbf{R}_k^r = \mathbf{I}_{3\times 3}$ and due to the rigid coupling $\mathbf{R}_k^c = \mathbf{I}_{3\times 3}$) equation (4.3) holds for any $\mathbf{\Delta R}^c$. For $\mathbf{R}_k^r \neq \mathbf{I}_{3\times 3}$, equation (4.3) imposes only two constraints on the orientation matrix. The angle about the rotation axis cannot be recovered. Furthermore, the matrix $\mathbf{R}_k^r - \mathbf{I}_{3\times 3}$ is singular and has rank two if $\mathbf{R}_k^r \neq \mathbf{I}_{3\times 3}$ [99]. Hence, for a known orientation matrix $\mathbf{\Delta R}^c$ equation (4.4) imposes up to two constraints on $\mathbf{\Delta t}^c$ and leaves one degree of freedom. The displacement along the rotation axis cannot be recovered. We further note that for pure translational motion the first term in equation (4.4) vanishes and the equation holds for any displacements. However, two constraints are imposed on $\mathbf{\Delta R}^c$ due to the alignment of the translation vectors.

4.2.2 The Ground Plane

In this thesis, the ground plane serves as a reference object for calibration. If the ground plane is observed by one or multiple cameras, additional constraints can be imposed on the parameters of the relative orientations and displacements between cameras as well as the parameters of the camera motion. In the following we introduce the mathematical relations between ground plane normals in multiple images as well as the relations between camera heights induced by the ground plane.

The relation between the observed ground plane normal in consecutive frames is given by

$$\mathbf{n}_{k+1}^c = \mathbf{R}_k^c \mathbf{n}_k^c. \tag{4.5}$$

Likewise, the relation between the normal in the coordinate frame of a camera $\mathcal{C}^c$ and the reference camera is given by

$$\mathbf{n}_k^r = \mathbf{\Delta R}^c \mathbf{n}_k^c. \tag{4.6}$$

The relation between the camera heights in consecutive frames is given by

$$h^c_{k+1} = h^c_k - (\mathbf{n}^c_{k+1})^T \mathbf{t}^c_k. \tag{4.7}$$

Correspondingly, the relation between the height of the reference camera and the height of a camera $\mathcal{C}^c$ is given by

$$h^c_k = h^r_k + (\mathbf{n}^r_k)^T \boldsymbol{\Delta} \mathbf{t}^c. \tag{4.8}$$

Notice that we can compute the height ratio of the camera centers in consecutive frames, h^c_{k+1}/h^c_k, using the results of the homography matrix decomposition, $\mathbf{R}^c_k$, $\mathbf{n}^c_k$, and $\mathbf{t}^c_k/h^c_k$. To this end, we divide equation (4.7) by h^c_k and propagate the ground plane normal using equation (4.5). In a similar manner the height ratio of the reference camera center and the camera center of $\mathcal{C}^c$, h^c_k/h^r_k, can be computed from equation (4.8).
Consequently, equations (4.5) to (4.8) enable propagating information about the height of camera centers over time and can be used to relate the height of all camera centers to the height of the reference camera.

4.2.3 Classes of Motion

In the following we define the different classes of motion that serve as the basis for our analysis. The motions resemble the typical driving maneuvers straight driving, turning, and driving on planar ground with and without rolling, pitching, and deflections. For the definition of planar and circular motion, i.e. turning, we make use of the instantaneous center of rotation as a means of motion parameterization. The definition of the instantaneous center of rotation can be found in the Appendix A.3. The definitions of the four classes of motion listed in the following are illustrated in Figure 4.3.

- Linear motion is the translation along a straight line without rotation, $\mathbf{R}^c_k = \mathbf{I}_{3\times3}$, $\mathbf{t}^c_{k+1} \times \mathbf{t}^c_k = \mathbf{0}_{3\times1}$, and $\|\mathbf{t}^r_k\|_2 = \|\mathbf{t}^c_k\|_2$. The translation direction vectors of each camera are aligned and in consequence, all cameras move at the same velocity. We further assume the translation to be parallel to the ground plane $(\mathbf{n}^c)^T \mathbf{t}^c_k = 0$. Linear motion resembles a straight driving maneuver and has only one degree of freedom, the velocity (non-uniform linear motion).

- Circular motion resembles a turning maneuver. It is the motion along the circumference of a circle, thus the instantaneous center of rotation is constant over time for each camera, respectively, $\mathbf{s}_k^c = \mathbf{s}^c$. Furthermore, the rotation axis direction coincides with the normal of the ground plane $\mathbf{r}_k^c = \mathbf{r}^c = \mathbf{n}^c$, and the translation is parallel to the ground plane $(\mathbf{n}^c)^T\mathbf{t}_k^c = 0$. Circular motion has also only one degree of freedom, the angular velocity (non-uniform circular motion).

- Planar motion is the translation in the plane parallel to the ground plane and the rotation about the ground plane normal $(\mathbf{n}^c)^T\mathbf{t}_k^c = 0$, and $\mathbf{r}_k^c = \mathbf{r}^c = \mathbf{n}^c$. In contrast to circular motion $\mathbf{s}_k^c$ is not constant over time. Planar motion has three degrees of freedom, namely the two parameters of the instantaneous center of rotation in the plane and the angular velocity.
- General motion is unconstrained and has the full six degrees of freedom.

In this thesis we do not make use of the constraints imposed by the non-holonomic motion of vehicles that adhere to the Ackermann steering principle. This type of motion would allow to describe planar motion using only two parameters but requires estimating the center and orientation of the rear axle.

4.2.4 Computation of Extrinsic Calibration Parameters

In the following we present the derivation of the observable parameters for each class of motion defined in Section 4.2.3. The derivations are based on the rigidity constraint between cameras (Section 4.2.1) and optionally the ground plane (Section 4.2.2). In addition, we distinguish between two different sensor outputs, a system of monocular cameras and a system of cameras that provide instantaneous depth measurements, e.g. stereo cameras.
Monocular systems suffer in general from the problem of scale ambiguity. Herein we tackle this problem by using either the ground plane as a reference object, an approach that uses at least image triplets, e.g. [72], or both. Next, we elaborate on the sensors and employed approaches and define the scale for datum definition for each of the three cases. Without loss of generality, we assume consecutive poses to be used starting at time index k.

- Pairwise evaluation of consecutive images in monocular sequences renders the propagation of velocity information impossible. Although it is in general possible to infer some information about the extrinsic calibration parameters, we restrict our analysis to the case of simultaneous observation of the ground plane. In this case the decomposition of the homography matrix (equation (3.14)) yields the scaled translation $\mathbf{t}_k^c/h_k^c$ which

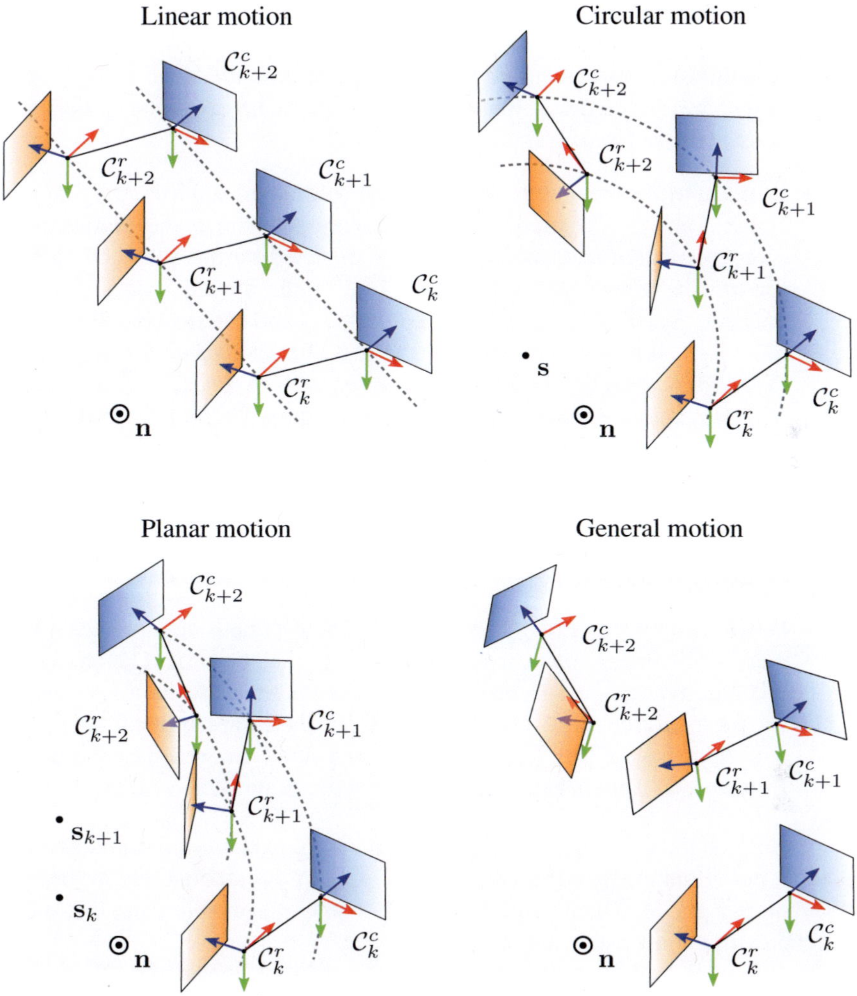

Figure 4.3: Schematic illustration of the four classes of motion that serve as the basis for our analysis. The classes are linear motion (translation along a straight line), circular motion (along the circumference of a circle), planar motion (motion in a plane), and unconstrained, general, motion.

serves as inputs. It was shown in Section 4.2.2 how relative velocity and camera height information can be propagated over time and between cameras by using the constraints imposed by the ground plane. Thus, for datum definition we define the scale by the camera height of the first camera pose of the reference camera h_k^r.

- If image triplets are used, 3D points can be triangulated from the first two cameras and then be used in the third camera to estimate the relative pose, a process called resectioning. This is depicted on the left hand side of Figure 4.4. Since we assume error free inputs we make no distinction between visual odometry approaches [84, 71] and classical bundle adjustment [98]. It is common to define the scale for each camera such that the distance between first two camera poses is equal to one, $\mathbf{t}_0^c/\lambda^c = 1$, where λ^c is a camera dependent scale factor. For datum definition we define the scale by the translation distance between the first two camera poses of the reference camera λ^r. Notice the resemblance in the datum definition between image triplets and pairwise evaluation of images. If the ground plane is observed, we adopt the datum definition of pairwise evaluation of consecutive frames.
- When using sensors that provide instantaneous depth measurements it is possible to directly recover the translation $\mathbf{t}_k^c$ and the correct height h_k^c. In this case no scale has to be defined for datum definition as it is provided by the measurements. This is depicted on the right hand side of Figure 4.4. Note that we treat a stereo camera as a single camera. The coordinate frame of the camera is associated with either of the stereo cameras.

Depending on the class of motion, sensor, and employed motion estimation approach, not all parameters can be observed. To express ambiguities we use the parameters τ and ω. The parameter τ denotes an unobservable scale factor and ω denotes an unobservable angle.
In the following, only two-camera systems ($C = 2$) are considered. The extension to multiple cameras is straightforward. For example, a three-camera system can be treated as two separate two-camera systems that share the reference camera. Furthermore, while incorporating multiple cameras might improve the robustness of the estimation in case of noisy observation, there is no difference in the cases examined here. In addition to the observability of parameters, the number of required consecutive poses is also of particular interest [100, 19]. In general, a low number of required consecutive poses is favorable. The results of

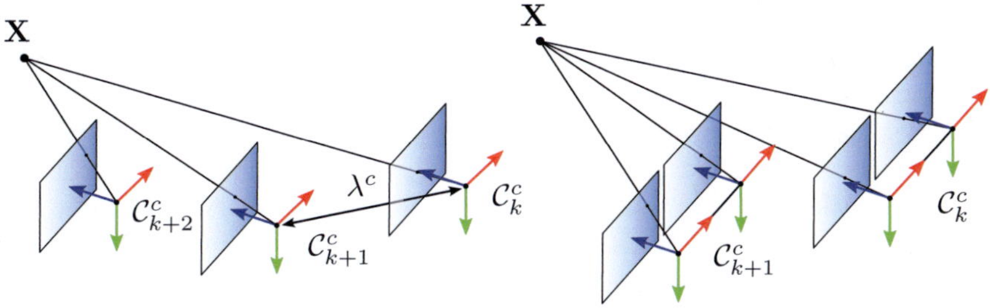

Figure 4.4: A 3D point reconstructed from image triplets (left) and a moving stereo camera with a triangulated 3D point (right). The translation distance between the first two frames is λ^c. The point triangulated from the corresponding cameras is used to propagate the scale information to the third camera pose by, e.g., resectioning. The stereo camera provides instantaneous depth measurements. Triangulated 3D points are used for motion estimation.

the derivations alongside with the minimum number of required poses are presented concisely in Table 4.1. Next we present the derivations ordered by classes of motion.

Linear Motion

Linear motion is the motion along a straight line. As there is no rotation, equation (4.3) holds for any choice of $\mathbf{\Delta R}^c$, and equation (4.4) simplifies to

$$\mathbf{t}^r_k = \mathbf{\Delta R}^c \mathbf{t}^c_k, \tag{4.9}$$

which imposes two constraints on the relative orientation. The rotation angle about the translation direction, however, cannot be observed. Thus, equation (4.9) holds for any relative camera orientation of the form $\mathbf{R}_{\mathbf{t}^r_k,\omega}\mathbf{\Delta R}^c$, where $\mathbf{R}_{\mathbf{t}^r_k,\omega}$ is a rotation about the (non-unit) rotation axis $\mathbf{t}^r_k$ with angle ω. The relative displacement cannot be recovered.
If the ground plane is observed, equation (4.6) can be employed. Due to the translational motion the observations of the plane normals are time independent

$$\mathbf{n}^r = \mathbf{\Delta R}^c \mathbf{n}^c. \tag{4.10}$$

By definition, the plane normal $\mathbf{n}^c$ is orthogonal to the translation vectors $\mathbf{t}^c_k$. Combined, equations (4.9) and (4.10) provide enough constraints to determine the relative orientation. To this end, we compose two auxiliary rotation matrices by constructing orthonormal right handed bases from the translation vectors and observed plane normals, $\mathbf{R}_{\mathbf{t}_k^r,\mathbf{n}^r}$ and $\mathbf{R}_{\mathbf{t}_k^c,\mathbf{n}^c}$. To construct the matrices we use Gram-Schmidt orthonormalization. This is explained in more detail in Appendix A.1. The two rotation matrices $\mathbf{R}_{\mathbf{t}_k^r,\mathbf{n}^r}$ and $\mathbf{R}_{\mathbf{t}_k^c,\mathbf{n}^c}$ represent the transformations from a translation vector and normal vector aligned coordinate frame into the camera coordinate frames. The relative orientation is then given by

$$\mathbf{\Delta R}^c = \mathbf{R}_{\mathbf{t}_k^r,\mathbf{n}^r}\mathbf{R}_{\mathbf{t}_k^c,\mathbf{n}^c}^T. \tag{4.11}$$

Note that only the translation direction is of interest here. In addition to the relative orientation, the relative height ratio h^c/h^r can be determined from the observations of the scaled translations $\mathbf{t}_k^c/h^c$ by enforcing $\|\mathbf{t}_k^r\|_2 = \|\mathbf{t}_k^c\|_2$. The correct camera heights can be measured directly if the sensor provides instantaneous depth measurements. To recover the observable parameters, only two consecutive poses are required.

Circular Motion

Circular motion is the motion along the circumference of a circle. The rotation axis directions of all cameras are aligned in the world and are time independent, yielding

$$\mathbf{r}^r = \mathbf{\Delta R}^c\mathbf{r}^c. \tag{4.12}$$

Equation (4.12) holds for any relative orientation $\mathbf{R}_{\mathbf{r}^r,\omega}\mathbf{\Delta R}^c$, thus the orientation about $\mathbf{r}^r$ remains ambiguous. In the case of circular motion, the observations of the ground plane normals do not provide additional information as they are aligned with the rotation axis directions.

If the translation velocities $\|\mathbf{t}_k^c\|_2$ are known, one can determine the circle radii

$$r^c = \frac{\|\mathbf{t}_k^c\|_2}{2\,\sin\left(\frac{\theta_k}{2}\right)}, \tag{4.13}$$

where θ_k is the (camera independent) angular velocity. If only scaled translations are observed, one can determine the radius to height, r^c/h^c, or radius to scale factor r^c/λ^c ratio. As in the case of linear motion, the observable parameters can be recovered from only two consecutive poses.

Planar Motion

Planar motion differs from circular motion in that the instantaneous center of rotation is time dependent. We make use of the property to compute the relative orientation between cameras by exploiting that the direction of the vectors $\mathbf{s}^c_{k+1} - \mathbf{s}^c_k$ are aligned (cf. Figure 4.3). Using the displacements $\mathbf{\Delta s}^c_k = \mathbf{s}^c_{k+1} - \mathbf{s}^c_k$, we construct auxiliary rotations matrices from $\mathbf{\Delta s}^c_k$ and $\mathbf{r}^c$ using Gram-Schmidt orthonormalization, $\mathbf{R}_{\mathbf{\Delta s}^c_k,\mathbf{r}^c}$. The relative orientations are then given by

$$\mathbf{\Delta R}^c = \mathbf{R}_{\mathbf{\Delta s}^r_k,\mathbf{r}^r}\, \mathbf{R}^T_{\mathbf{\Delta s}^c_k,\mathbf{r}^c}. \tag{4.14}$$

Note that $\mathbf{\Delta s}^c_k$ and the rotation axis are orthogonal and only the directions of $\mathbf{\Delta s}^c_k$ are of interest. If at least image triplets or the ground plane are used one obtains $\mathbf{\Delta s}^c_k/\lambda^c$ or $\mathbf{\Delta s}^c_k/h^c$, respectively. Next, we derive the camera displacements. We divide equation (4.4) by λ^rwhich leads to

$$(\mathbf{R}^r_k - \mathbf{I}_{3\times3})\,\frac{\mathbf{\Delta t}^c}{\lambda^r} - \mathbf{\Delta R}^c\,\frac{\mathbf{t}^c_k}{\lambda^c}\frac{\lambda^c}{\lambda^r} + \frac{\mathbf{t}^r_k}{\lambda^r} = \mathbf{0}, \tag{4.15}$$

such that all translations appear normalized. We stack the equations of two consecutive motions and rewrite the result in form of a linear system of equations

$$\underbrace{\begin{bmatrix} \mathbf{R}^r_k - \mathbf{I}_{3\times3} & -\mathbf{\Delta R}^c\mathbf{t}^c_k/\lambda^c \\ \mathbf{R}^r_{k+1} - \mathbf{I}_{3\times3} & -\mathbf{\Delta R}^c\mathbf{t}^c_{k+1}/\lambda^c \end{bmatrix}}_{=:\mathbf{A}^c_k} \begin{pmatrix} \mathbf{\Delta t}^c/\lambda^r \\ \lambda^c/\lambda^r \end{pmatrix} = - \begin{pmatrix} \mathbf{t}^r_k/\lambda^r \\ \mathbf{t}^r_{k+1}/\lambda^r \end{pmatrix}. \tag{4.16}$$

For planar motion the matrices $\mathbf{A}^c_k$ are rank-deficient and do not constrain the nonplanar parts of the displacement vectors. In other words, the linear equation systems (4.16) hold for any scaled displacement vectors of the form $\mathbf{\Delta t}^c/\lambda^r + \tau^c\mathbf{r}^r$, with $\tau^c \in \mathbb{R}$. Without using the ground plane we cannot determine the nonplanar part of the displacements, $\mathbf{\Delta t}^c_\perp$. The planar part $\mathbf{\Delta t}^c_\parallel$ can be computed by augmenting equation (4.16) by

$$0 = (\mathbf{r}^r)^T\,\mathbf{\Delta t}^c = (\mathbf{r}^r)^T\,(\mathbf{\Delta t}^c_\parallel + \mathbf{\Delta t}^c_\perp), \tag{4.17}$$

thus enforcing $\mathbf{\Delta t}^c_\perp = \mathbf{0}$.

If the ground plane is used we can substitute h^c and h^r for λ^c and λ^r. The nonplanar part of the camera displacements can then be computed directly from

$$\mathbf{\Delta t}^c_\perp / h^r = \mathbf{n}^r \left(h^c / h^r - 1 \right) \tag{4.18}$$

(cf. equation (4.8)), where h^c/h^r is obtained from the linear equation system (4.16).
In case of planar motion all parameters can be recovered if the ground plane is observed. At least three consecutive poses or correspondingly two consecutive motions are required, respectively. This comes at no surprise as using only one motion is equivalent to the case of circular motion. The special case of combining linear and circular motion is not covered.

General Motion

General motion has six degrees of freedom and is unconstrained. Metric calibration is possible in all considered cases. The relative orientations can be determined by constructing auxiliary rotation matrices from the time dependent rotation axis directions

$$\mathbf{\Delta R}^c = \mathbf{R}_{\mathbf{r}^r_{k+1}, \mathbf{r}^r_k} \, \mathbf{R}^T_{\mathbf{r}^c_{k+1}, \mathbf{r}^c_k} . \tag{4.19}$$

The camera displacement can be determined by solving the linear equation system (4.16). The matrix $\mathbf{A}^c_k$ is not rank-deficient in the case of general motion. This approach requires two motions or correspondingly three consecutive poses, respectively. However, if the ground plane is observed only two consecutive poses or one motion is required, respectively.
The relative camera orientations can be determined from the ground plane normals and rotation axis directions as

$$\mathbf{\Delta R}^c = \mathbf{R}_{\mathbf{r}^r_k, \mathbf{n}^r_k} \, \mathbf{R}^T_{\mathbf{r}^c_k, \mathbf{n}^c_k} . \tag{4.20}$$

For the derivation of the camera displacements we substitute h^c_k and h^r_k for λ^c and λ^r in equation (4.15), respectively, rendering the equation time dependent.

We then use equation (4.8) to substitute h_k^c/h_k^r by $1 + (\mathbf{n}_k^r)^T \mathbf{\Delta t}^c/h_k^c$, which yields, after rearranging

$$\underbrace{\left[\mathbf{R}_k^r - \mathbf{I}_{3\times 3} - \mathbf{\Delta R}^c \frac{\mathbf{t}_k^c}{h_k^c} (\mathbf{n}_k^r)^T\right]}_{=:\mathbf{B}_k^c} \frac{\mathbf{\Delta t}^c}{h_k^r} = \mathbf{\Delta R}^c \frac{\mathbf{t}_k^c}{h_k^c} - \frac{\mathbf{t}_k^r}{h_k^r}. \tag{4.21}$$

The matrix $\mathbf{B}_k^c$ has in general full rank. Hence, $\mathbf{\Delta t}^c/\lambda^r$ can be computed by solving the linear equation system.

4.2.5 Summary

We have derived algorithms to determine the observable extrinsic calibration parameters for each of four different classes of motion and different sensor outputs as well as different approaches for motion estimation. The results are presented concisely in Table 4.1. We observe that neither pure translation nor pure circular motion provide enough information to recover the extrinsic calibration, independent of the algorithm input. For planar motion, the ground plane is required as a reference object to enable metric calibration. For general motion, metric calibration is always possible in general. Interestingly, by using the ground plane, only two consecutive poses are required. When using the ground plane, image triplets do not provide additional information. However, when using noisy observations, this approach is likely to outperform pairwise evaluation of images.

	Linear motion		Circular motion	
Image triplets	$\mathbf{R}_{\mathbf{t}_k^r,\omega^c}\Delta\mathbf{R}^c$ -	(2)	$\mathbf{R}_{\mathbf{r}^r,\omega^c}\Delta\mathbf{R}^c$ -	(2)
Instantaneous depth	$\mathbf{R}_{\mathbf{t}_k^r,\omega^c}\Delta\mathbf{R}^c$ -	(2)	$\mathbf{R}_{\mathbf{r}^r,\omega^c}\Delta\mathbf{R}^c$ r^c	(2)
Pairwise evaluation/ image triplets with ground plane	$\Delta\mathbf{R}^c$ h^c/h^r	(2)	$\mathbf{R}_{\mathbf{r}^r,\omega^c}\Delta\mathbf{R}^c$ r^c/h^c	(2)
Instantaneous depth with ground plane	$\Delta\mathbf{R}^c$ h^c	(2)	$\mathbf{R}_{\mathbf{r}^r,\omega^c}\Delta\mathbf{R}^c$ h^c, r^c	(2)

	Planar motion		General motion	
Image triplets	$\Delta\mathbf{R}^c$ $\Delta\mathbf{t}^c/\lambda^r + \tau^c\mathbf{r}^r$	(3)	$\Delta\mathbf{R}^c$ $\lambda^r\Delta\mathbf{t}^c$	(3)
Instantaneous depth	$\Delta\mathbf{R}^c$ $\Delta\mathbf{t}^c + \tau^c\mathbf{r}^c$	(3)	$\Delta\mathbf{R}^c$ $\Delta\mathbf{t}^c$	(3)
Pairwise evaluation/ image triplets with ground plane	$\Delta\mathbf{R}^c$ $\Delta\mathbf{t}^c/h^r$	(3)	$\Delta\mathbf{R}^c$ $\Delta\mathbf{t}^c/h_k^r$	(2)
Instantaneous depth with ground plane	$\Delta\mathbf{R}^c$ $\Delta\mathbf{t}^c$	(3)	$\Delta\mathbf{R}^c$ $\Delta\mathbf{t}^c$	(2)

Table 4.1: Observable relative orientations and displacements for different classes of motion and sensor outputs as well as algorithms. The results for the relative orientations are shown on the top, the results for the camera displacements are shown on the bottom, respectively. The scalars τ and ω denote an unobservable scale and an unboservable angle, respectively. The minimum number of required consecutive poses is shown in parentheses in blue, respectively.

4.3 Calibration from Overlapping Fields of View

In the remainder of this chapter we elaborate on the extrinsic calibration of a multi-camera system using pairwise overlapping fields of view.
If the fields of view of two cameras overlap and corresponding image points can be established the relative orientation and displacement direction, comprised by the essential matrix (cf. Section 3.3.2), can be estimated. However, the scale of the displacement, i.e. the baseline, cannot be recovered. In a multi-camera system several overlapping fields of view may exist. If certain conditions are met, metric calibration based on the epipolar constraint is possible. In the following these conditions are elucidated. To this end, the multi-camera system and the relative pose transformations are represented as a graph. We make use of established definitions of graph theory to formulate two necessary conditions which, if met, allow to apply a matrix rank test which yields a binary measure of the observability of the (metric) extrinsic calibration.
Graphs are commonly used in computer vision as a means to model mathematical problems (e.g. [97]) and in particular in mutli-camera calibration (e.g. [5]). We represent the mutli-camera system as an undirected simple graph $\mathcal{G} = (\mathcal{V}, \mathcal{E})$ [78], where a camera is represented by a vertex $c \in \mathcal{V}$, where $\mathcal{V}$ is the set of vertices. The set of edges is $\mathcal{E}$. If two cameras have overlapping fields of view and the relative orientation and displacement direction can be estimated we call the cameras adjacent and they are joined by an edge. In the following only multi-camera systems with more than two cameras are considered ($C > 2$). A two camera system, as stated above, can be calibrated metrically if the essential matrix can be estimated. This case can be regarded as a special, trivial case.
A multi-camera system can be calibrated metrically from pairwise overlapping fields of view only if the two following necessary conditions hold.

- The graph has to be connected, i.e. any two cameras are linked by a sequence of pairwise adjacent cameras.
- All edges have to be contained in at least one simple cycle, i.e. a sequence of adjacent vertices starting and ending at the same vertex without repetitions of vertices and edges (except for the first and last vertex).

The first condition ensures that the relative orientation between all cameras can be derived. If the graph is connected, the relative orientation between any two cameras c and d can be computed by following the path[2] from camera $\mathcal{C}^c$ to

[2] We define a path as the ordered sequence of edges.

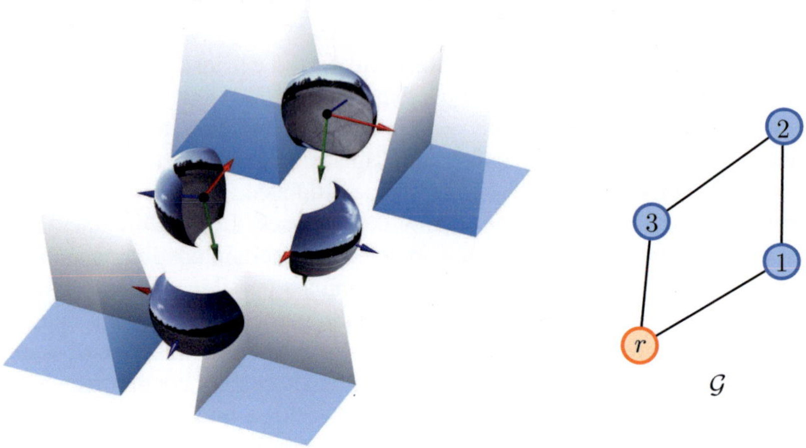

Figure 4.5: Four-camera system with overlapping fields of view (left) and corresponding graph representation $\mathcal{G}$ (right). To illustrate the fields of view of fisheye cameras image data is projected onto spherical sectors. The boundaries of the overlapping fields of view are indicated by blue patches on the ground plane and grey planes elsewhere. The corresponding graph representation is shown on the right. The orange circle represents the reference camera ($r = 0$) and blue circles represent the remaining cameras. The edges indicate overlapping fields of view from which the relative orientation and displacement direction can be estimated, respectively.

$\mathcal{C}^d$ and concatenating the relative orientation matrices corresponding to the traversed edges. In particular, the relative orientation with respect to the reference camera, $\mathbf{\Delta R}^c$, $c \in \{0, ..., C-1\}$, can be computed.
The concatenation of relative poses along a circular path in the graph has to yield an identity matrix. The second conditions ensures that this constraint can be imposed on all edges. Figure 4.5 illustrates the overlapping fields of view and the corresponding graph representation of the four-camera system used for experimental evaluation. The graph shown in the figure is a cycle graph and therefore meets both conditions. However, if all cameras centers were aligned the baselines could not be recovered[3]. To identify such configurations we apply a matrix rank test.

[3] Only pairwise camera constraints are considered.

To this end, first the relative displacement directions are computed and transformed into the reference coordinate frame. For example, the relative displacement direction from camera $\mathcal{C}^1$ to camera $\mathcal{C}^2$ in the reference coordinate frame is given by $\left(\mathbf{\Delta t}^2 - \mathbf{\Delta t}^1\right) / \|\mathbf{\Delta t}^2 - \mathbf{\Delta t}^1\|_2$. The individual displacements $\mathbf{\Delta t}^c$ with respect to the reference camera are unknown, but the relative displacement directions can be computed directly using the (known) relative orientation and decomposition of the essential matrix.
Next, for each simple cycle in the graph, a matrix is constructed by appending the displacement direction vectors corresponding to the edges in the cycle. Without loss of generality, the signs of the displacement directions vectors are chosen such that they are always pointing at the (camera) vertex with the higher index. The matrices are then stacked together such that each column corresponds to one edge. Direction vectors of edges not contained in simple cycle are set to zero. The multi-camera system can be calibrated metrically if and only if the rank of the matrix is equal to the number of edges minus one. For the example in Figure 4.5, we form the homogeneous system of equations

$$\left[\frac{\mathbf{\Delta t}^1}{\|\mathbf{\Delta t}^1\|_2}, \quad \frac{\mathbf{\Delta t}^2-\mathbf{\Delta t}^1}{\|\mathbf{\Delta t}^2-\mathbf{\Delta t}^1\|_2}, \quad \frac{\mathbf{\Delta t}^3-\mathbf{\Delta t}^2}{\|\mathbf{\Delta t}^3-\mathbf{\Delta t}^2\|_2}, \quad \frac{\mathbf{\Delta t}^3}{\|\mathbf{\Delta t}^3\|_2}\right]\boldsymbol{\lambda} = \mathbf{0}_{3\times 1}. \tag{4.22}$$

The matrix is of size 3×4 and contains only one simple cycle. Recall that $\mathbf{\Delta t}^r = \mathbf{0}_{3\times 1}$. The four-camera system can thus be calibrated metrically if the rank of the matrix is three. By enforcing $\boldsymbol{\lambda}^T\boldsymbol{\lambda} = 1$ we obtain the non-trivial solution of the equation system which yields the vector of baselines $\boldsymbol{\lambda}$. Instead of using all simple cycles in a graph it is sufficient to only consider the elements of a cycle basis.
The matrix rank test can only be applied to error free data and is thus of little relevance in practice. However, it can be used as a means to identify singular configurations. E.g. the matrix is of rank two if the four camera centers are coplanar and of rank one if the camera centers are collinear. Furthermore, a cycle graph of length five cannot be calibrated metrically since the corresponding matrix can have rank three at most.
The camera centers of the system used in the experimental evaluation are not coplanar. An offline extrinsic calibration approach based on overlapping fields of view using a similar system for evaluation is presented in our previous work [43].

5 Establishing Point Correspondences

Establishing image point correspondences is fundamental to our calibration approach. The correspondences are used to estimate the relative pose between adjacent cameras, camera motion, and the ground plane.
Large spatial camera displacements accompanied by severe lens distortions render the detection of putative correspondences difficult. This holds in particular in close proximity, e.g. within the order of magnitude of the baseline. Correspondences in this range are of particular interest to calibration as they allow for a more accurate estimation of the camera displacement.
Therefore, we propose warping the images prior to extracting feature correspondences to establish image similarity. To this end, we approximate the scene by the ground plane in close proximity and infinitely distant objects elsewhere. This approach is applied to both, cameras that are either offset spatially as in a stereo setup or both spatially and temporarily as it is the case for a moving monocular camera. This allows treating both cases uniformly.
Earlier versions of the work presented in this chapter have been published in [43] and [45].

5.1 Wide Baseline Matching

The literature offers a large variety of different methods to establish image point correspondences. In the following we give a brief overview and highlight work relevant to the problem of wide baseline matching. Typically, the task of establishing image correspondences is divided into three steps. First, distinctive points or regions are detected in the image, e.g. corners or blobs. A descriptor is then used to capture local image properties and the information is stored in a feature vector. Finally, the feature vectors are matched across images to establish putative correspondences. The methods are commonly classified by their invariance with respect to different image transformations, such as spatial or range transformations, and their computational complexity [33, 65].
Invariance against certain spatial transformations can be achieved by normalizing image regions prior to extracting the feature vector. The transformation normalizing the image region can, e.g., be derived from an analysis of the sec-

ond moment matrix of intensity gradients from the dominant gradient direction [65, 56]. Invariance of the feature descriptor is not required.
A different approach was presented by Morel and Yu [68]. Instead of normalizing image regions, multiple different viewpoints are simulated by distorting the images accordingly. All distorted image regions are then compared using the scale invariant feature transform (SIFT) [56]. The distortion caused by the simulated viewpoints is approximated by an affine transformation. The method is thus termed affine SIFT (ASIFT). Through various techniques it is possible to achieve invariance to shifts (e.g. [79]), Euclidean transformations (e.g. [80]), similarity transformations (e.g. [56, 52]), affine transformations (e.g. [68]), and projective transformations [11] to some extent.
In addition, feature detectors and descriptors have been proposed that account for the geometric image distortions introduced by wide-angle lenses (e.g. [102]). However, these algorithms do consider the perspective distortions caused by large viewpoint variations. Figure 5.1 exemplary shows two images captured simultaneously from a right and rear-facing camera mounted on our test vehicle. Three magnified corresponding image patches are shown. While the treetop is similar in appearance, the ground in front of the parked automobile is skewed and the parking spot markings are significantly distorted.
To quantify the image distortion, we employ the metric proposed by Morel and Yu [68]. The geometric image distortions are approximated by an affine transformation. The transition tilt corresponds to the ratio of eigenvalues of the upper left two by two affine transformation matrix. Geometrically, it corresponds to change of the aspect ratio of a rotated window. In the example shown in Figure 5.1 the transition tile is approximately 2.4 below the parked vehicle and 7 on the parking spot marking[1].
ASIFT has been shown to work under severe distortions. Applying it to this example, we were able to establish some correspondences in the most distinctive regions such as on the line markings, but the method failed in case of their absence. For this reason, and due to the high complexity of ASIFT we approach the problem differently.
We use estimates of the current camera configuration and ground plane to warp images in order to compensate image distortions between two views. To this end, we apply a coarse approximation of the scene. This allows using feature detectors and descriptors which are not invariant to geometric distortions but have a significantly lower computational complexity. The employed algorithms

[1] Ground plane and camera poses are known from a reference calibration.

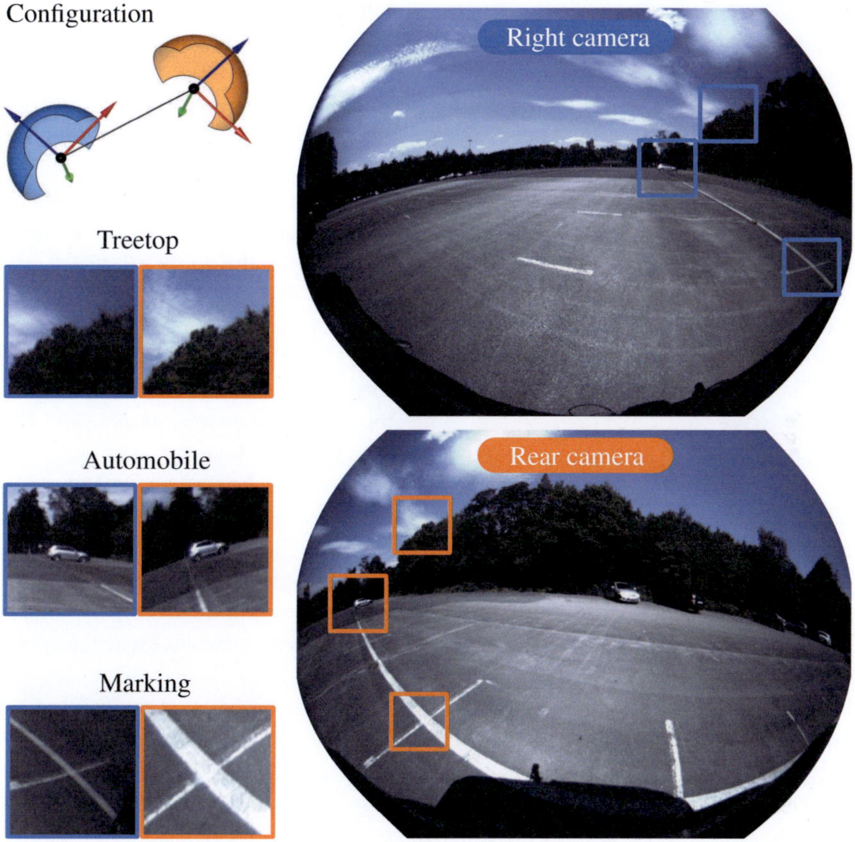

Figure 5.1: Two images captured simultaneously from a right and rear-facing camera mounted on our test vehicle are shown (right). The three orange and blue squares indicate the position of the magnified corresponding patches (treetop, car, marking). The configuration of the cameras is shown in the top left.

are the features from accelerated segment test (FAST) feature detector by Rosten and Drummond [79] and the binary robust independent elementary features (BRIEF) feature descriptor by Calonder et al. [13].

5.2 Scene Geometry Approximation

To compensate for the strong distortions between corresponding image regions our goal is to find an image mapping that allows warping one image into the other such that these regions coincide. Warping is applied prior to the feature detection and extraction.

In unobstructed image regions the mapping relating both images is defined by the camera configuration, the geometric imaging characteristics of the camera, and the 3D scene geometry. The scene geometry is not known a-priori and its estimation is not within the scope of this thesis. Hence, we approximate the scene by geometric primitives. This approach was proposed for the ground plane in the context of stereo vision by Burt et al. [12], where equi-disparity on the road is obtained by applying a linear transformation to a stereo rectified image pair [31].We adopt this concept and apply it to spatially as well as spatially and temporarily offset cameras. Objects above the ground plane are assumed to be infinitely far away. Thus, the scene is approximated by the ground plane in close proximity and infinitely distant objects elsewhere. Stereo rectification is not applied.

In the following we derive the mappings for image warping. The notation differs slightly from the previous chapters as spatially and temporarily offset cameras are treated in a unified way. For this reason, time indices were omitted. However, the mappings are time dependent in general.

Let the relative pose transformation $\mathbf{\Delta T}$ between two cameras, the ground plane normal $\mathbf{n}$ in the coordinate system of the first camera, and the corresponding height h be given. The transformation of a 3D point $\mathbf{X}$ from the first into the second camera coordinate system is then given by $\mathbf{X}' = \mathbf{\Delta R X} + \mathbf{\Delta t}$ (cf. Chapter 3). If the point is located on the ground plane the relation between the corresponding rays $\mathbf{x} \leftrightarrow \mathbf{x}'_g$ is given by the homography

$$\mathbf{x}'_g = \left(\mathbf{\Delta R} - \frac{\mathbf{\Delta t n}^{\mathrm{T}}}{h} \right) \mathbf{x} = \mathbf{H x}. \tag{5.1}$$

The mapping relating the image points also has to take the nonlinear projection onto the image plane into account. We write the image to image mapping as $\mathbf{u}'_g = \boldsymbol{\kappa} \left(\mathbf{H}\, \boldsymbol{\kappa}^{-1} (\mathbf{u}) \right) = \boldsymbol{\Psi}_g (\mathbf{u})$.

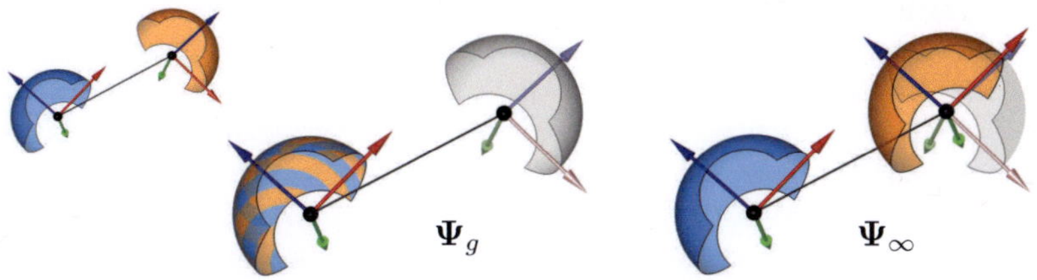

Figure 5.2: Illustration of the mappings $\boldsymbol{\Psi}_g$ and $\boldsymbol{\Psi}_\infty$. The original camera setup is shown in the upper left corner. The image captured by the first camera (orange) is warped into the image captured by the second camera (blue). When applying the ground plane induced mapping, the warped image appears as being captured from the viewpoint of the second camera. The mapping via the infinite homography corresponds to a rotation of the camera accompanied by image distortions due to the different intrinsic parameters of the cameras (right).

For infinitely distant objects we apply an infinite homography [31], i.e. for $h \to \infty$ the second term in equation (5.1) vanishes. The transformation simplifies to a rotation

$$\mathbf{x}'_\infty = \boldsymbol{\Delta}\mathbf{R}\mathbf{x}, \tag{5.2}$$

and the corresponding mapping between the image points is given by $\mathbf{u}'_\infty = \boldsymbol{\kappa}\left(\boldsymbol{\Delta}\mathbf{R}\,\boldsymbol{\kappa}^{-1}(\mathbf{u})\right) = \boldsymbol{\Psi}_\infty(\mathbf{u})$. Figure 5.2 illustrates the two mappings.

5.3 Image Resampling and Smoothing

Image resampling is applied to warp the image I captured by the first camera into the image I' captured by the second camera. Without loss of generality, we define the first image to be the source image and the second image to be the target image. Image resampling requires filtering and subsequent sampling. Ideally, filtering and sampling would not introduce aliasing or blur. However, magnified image regions are inevitably missing higher spatial frequency content. For this reason, we propose applying an appropriate smoothing filter to the target image so that both images exhibit the same local smoothness. In the remainder of this section, we will present the resampling filter and appropriate smoothing filter.
To resample the image, we apply the method published by Heckbert [32] which is summarized briefly in the following. Ideal resampling consists of the following four stages [88]:

1. The continuous image is reconstructed using image interpolation.
2. The result is (forward) warped according to the mapping.
3. Pre-filtering is applied to band-limit the signal.
4. The output is sampled at integer positions.

It was shown in [32] that these four stages can be rearranged and combined into a single filter that works on sampled positions only. To this end, the sampling grid of the target image is warped backwards into the source domain, yielding the resampling grid. The resampling grid does not coincide with the sampling grid of the source image in general, thus interpolation is required. Instead of applying the pre-filter to the warped image, the filter is also warped backwards into the source domain. The interpolation filter and warped pre-filter are then combined into a single filter. This step works for linear filters such as Gaussian filters which are closed under convolution. The resulting filter is space variant in general due to the space variant mapping.
In many applications, Gaussian filters are unpopular as they introduce significant blur [95]. Greisen et al. [27] present an approach based on the work of Heckbert that reduces the blurring by careful adjustment of the filter parameters. However, it was shown by Calonder et al. [13] that smoothing prior to feature extraction yields better results. Furthermore, the rapidly diminishing tails of the Gaussian function allow for truncation without introducing significant aliasing. For these reasons, we employ Gaussian filters for reconstruction and pre-filtering.

The reconstruction and pre-filter have identical covariance matrices $\boldsymbol{\Sigma}$ [32]. A point $\mathbf{u}'$ on the sampling grid in the target image and the corresponding point $\mathbf{u}_r$ on the resampling grid in the source image are related by $\mathbf{u}' = \boldsymbol{\Psi}(\mathbf{u}_r)$ (cf. section 5.2). The mapping is then linearized around $\mathbf{u}_r$, $\boldsymbol{\Psi}(\mathbf{u}_{r0} + \Delta\mathbf{u}_r) \approx \mathbf{J}\Delta\mathbf{u}_r + \boldsymbol{\Psi}(\mathbf{u}_{r0})$, where $\mathbf{J}$ is the Jacobian matrix evaluated at $\mathbf{u}_{r0}$. The covariance matrix of the Gaussian filter applied to the source image is then

$$\boldsymbol{\Sigma}_I = \boldsymbol{\Sigma} + \mathbf{J}^{-1}\boldsymbol{\Sigma}\mathbf{J}^{-\mathrm{T}}, \tag{5.3}$$

i.e. the convolution of the reconstruction filter and inversely transformed pre-filter. We transform the covariance matrix into the target domain

$$\boldsymbol{\Sigma}_{I'} = \mathbf{J}\boldsymbol{\Sigma}_I\mathbf{J}^T = \mathbf{J}\boldsymbol{\Sigma}\mathbf{J}^T + \boldsymbol{\Sigma}, \tag{5.4}$$

to obtain the corresponding smoothing filter applied to image I'. Since both filters are applied before feature extraction and matching this processing step is termed pre-warping and smoothing. Figure 5.3 shows the pre-warped and smoothed image regions corresponding to Figure 5.1. Since the two mappings only coincide for points at infinity, a distinctive image discontinuity can be observed at the boundary between the pre-warped regions. Therefore, the parameters are chosen such that the regions overlap during processing. Figure 5.4 shows a close-up of another scene. Note that both images are significantly blurred but appear similar in regions where the ground plane assumption holds.

Figure 5.3: An exemplary result of the pre-warping and smoothing step is shown. The smoothed image is shown on blue and the pre-warped image is shown in orange. The applied mapping is the infinite homography in the upper, and the ground plane induced homography in the lower image part, respectively. The data was captured during a turning maneuver. Due to the rolling of the vehicle, image regions on the ground plane do not coincide perfectly. Distant features (e.g. clouds) coincide.

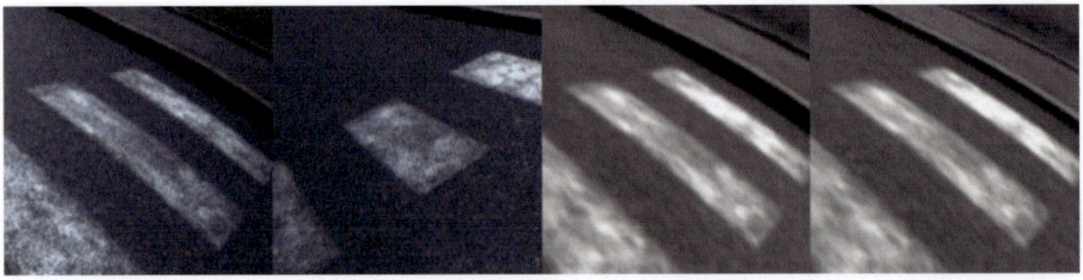

Figure 5.4: Another exemplary result of the image pre-warping and smoothing step is shown. The left two images show the corresponding cut-outs of a crosswalk captured by a front and right-facing camera. The right two images show the smoothed and pre-warped image, respectively. Note that structures off the ground plane like the curbstones do not appear similar due to the violation of the ground plane assumption.

6 Robust Homography Estimation

In this chapter, we present a robust method for estimating the frame-to-frame homography induced by the ground plane. The ground plane plays a fundamental role for the calibration approach presented in this thesis. On the one hand, it is used to recover the scale in successive frames and to constrain the motion model, on the other hand it serves as a reference object observed by all cameras simultaneously (Chapter 4.2). Furthermore, we employ the ground plane induced homography to establish image correspondences in successive frames and between cameras (Chapter 5). The homography matrix comprises motion as well as ground plane information. Still, homography estimation is challenging, as measurements are often not only corrupted by sparse gross outliers, but might also contain other structures, which are inconsistent with the ground plane, such as curbstones and sidewalks. Several well studied algorithms regarding the identification of sparse gross outliers have been proposed in the past, with random sample consensus (RANSAC) [21, 31] being the most prominent one. However, identifying structural outliers remains a challenging problem due the outliers' inner coherence which can cause strong systematic errors [91]. In homography and plane estimation structural outliers often cause plane fits that do not correspond to any physical plane in the scene. This becomes particularly challenging in the presence of planes with similar parameters, e.g. the road plane and a slightly elevated sidewalk plane in a street. To circumvent this problem, approaches estimating multiple structures simultaneously can be employed (e.g. [16, 96]). The high complexity of these algorithms, attributed to the fact that the number of structures, structure parameters, and noise levels have to be estimated and adjusted concurrently, makes them impractical for applications in the context of real-time applications. Under the assumption that information about the structure of interest is provided initially, the task can be simplified to robust tracking. Several approaches adopting this concept have been proposed hitherto [42, 90, 4, 104, 55]. Yet, none of them has been designed to work in scenarios where the observed scene is dominated by structural outliers, a situation typically encountered when attaching cameras to the side of a vehicle.

The method presented in this chapter relies on an initial estimate of the motion and ground plane parameters. From a statistical analysis on feature point correspondences local adaptive thresholds are derived that comply with a predefined expected false positive rate criterion. To this end, the positions of feature points in successive view are predicted and compared to hypothetical positions of the feature points induced by planes parallel to the ground plane. The false positive rate refers to the probability of incorrectly identifying the hypothetical feature point on the virtual plane as an inlier.

In the following the planar parallax decomposition which will be used throughout this chapter is introduced. Then, the derivation of the threshold values and the acceptance region is presented. Finally, we show how the risk of rejecting inliers can be mitigated by employing a sequential processing scheme, and how this scheme can be embedded into a Kalman filter. Results are shown for a sequence captured in the inner city, and compared to a RANSAC-based approach. The method presented here differs from the preliminary work [43] by incorporating the uncertainty of the motion and ground plane prediction and by considering non-isotropic Gaussian noise.

6.1 Planar Parallax Decomposition

We can interpret the displacement of positions of corresponding features points in successive camera views as a motion field in the image. For a static scene and moving camera, the motion field can be decomposed into the motion field of an arbitrary physical or virtual plane and a residual parallax field [47, 83]. The motion field induced by a plane can be described by a homography. The residual (planar) parallax field is an epipolar field, i.e. all vectors point towards the epipole[1]. In the following this is explained in more detail.

We consider the situation depicted in Figure 6.1. A ray $\mathbf{x}$ originating from the first camera center intersects an object in $\mathbf{X}_o$, and the ground plane in $\mathbf{X}_g$. The corresponding rays in the second view are $\mathbf{x}'_o$ and $\mathbf{x}'_g$, respectively. If the reference plane coincides with the ground plane, a pair of corresponding rays $\mathbf{x} \leftrightarrow \mathbf{x}'_g$ is related by the ground plane induced homography (cf. Chapter 3)

$$\mathbf{x}'_g = \mathbf{H}_k \mathbf{x} = \left(\mathbf{R}_k - \frac{\mathbf{t}_k {\mathbf{n}_k}^T}{h_k} \right) \mathbf{x}, \qquad (6.1)$$

[1] Or from the epipole outwards depending on the direction of the camera motion.

whereas the relation $\mathbf{x} \leftrightarrow \mathbf{x}'_o$ is given by

$$\mathbf{x}'_o = \widetilde{\mathbf{H}}_k \mathbf{x} = \left(\mathbf{R}_k - \frac{\mathbf{t}_k {\mathbf{n}_k}^T}{h_k + \Delta h} \right) \mathbf{x}, \tag{6.2}$$

with $\Delta h = \mathbf{n}_k^T (\mathbf{X}_o - \mathbf{X}_g)$ being the height difference with respect to the ground plane. It follows that every static 3D point $\mathbf{X}$ can be considered as being transformed into the coordinate system of the successive view by a homography using a virtual plane containing $\mathbf{X}$, which is parallel to the ground plane. The transformation between $\mathbf{x}'_g$ and $\mathbf{x}'_o$ is given by

$$\mathbf{x}'_o = \widetilde{\mathbf{H}}_k \mathbf{H}_k^{-1} \mathbf{x}'_g = \left(\mathbf{I}_{3\times 3} + \frac{\mathbf{t}_k \mathbf{n}_{k+1}^T}{h_{k+1}} \left(\frac{\Delta h}{h_k + \Delta h} \right) \right) \mathbf{x}'_g. \tag{6.3}$$

The complete derivation is given in Appendix A.4. It can easily be verified from equation (6.3) that the planar parallax is an epipolar field, as $\mathbf{x}'_o$ is a linear combination of $\mathbf{x}'_g$ and $\mathbf{t}_k$, and the image of $\mathbf{t}_k$ is the epipole [31]. The planar parallax vector in the image is given by $\mathbf{d} = \boldsymbol{\kappa}(\mathbf{x}'_o) - \boldsymbol{\kappa}(\mathbf{x}'_g) = \mathbf{u}'_o - \mathbf{u}'_g$, where $\boldsymbol{\kappa}(\cdot)$ is the projection onto the image plane. Figure 6.2 exemplarily shows a sparse parallax field superimposed on an image of a backward-facing camera. While planar parallax vectors on the road surface are small and mainly caused by noise corruption, planar parallax vectors on the sidewalk (left) show a predominant direction and length.

6.2 Local Adaptive Thresholds

Planar parallax gives strong cues towards the identification of points to the ground plane. However, this only holds in close proximity to the camera. In the distance planar parallax vanishes (e.g. see center region in Figure 6.2). Moreover, obtained parallax vectors are subject to noise and a reliable ground plane and motion prediction might not always be given. For this reason, acceptance regions and associated thresholds for inlier identification should adapt over time with respect to the prediction uncertainty and be local to account for the expected parallax in the respective image region. In the following we present a statistical analysis of the expected position of corresponding features in the image and the expected parallax for corresponding features not located on the ground plane. From this we derive local adaptive thresholds that are based on a false positive criterion.

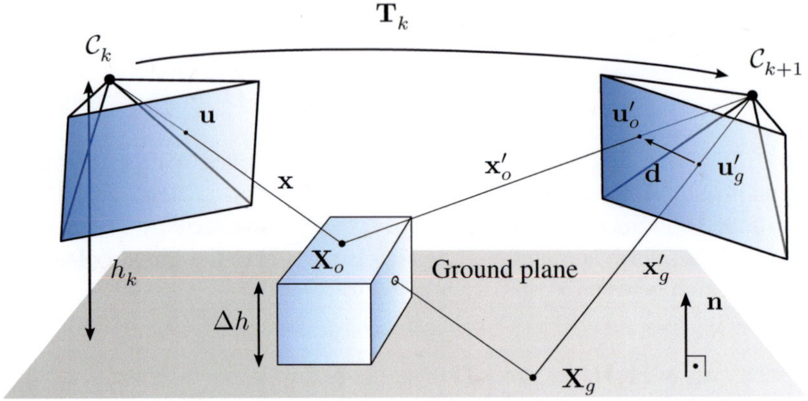

Figure 6.1: Schematic illustration of the considered example. A point in the first view is associated with two points in the second view, on an object, and on the ground plane.

We assume a given prediction of the current motion and ground plane parameters $\widehat{\boldsymbol{\xi}}$ along with the associated covariance matrix $\mathbf{P}$, a pair of corresponding image points in successive images $\mathbf{u} \leftrightarrow \mathbf{u}'$, and the associated position covariance matrix $\boldsymbol{\Sigma}$. The predicted image position in the successive view is then given by

$$\widehat{\mathbf{u}}'_g = \mathbf{h}\left(\mathbf{u}, \widehat{\boldsymbol{\xi}}\right), \tag{6.4}$$

assuming the corresponding 3D point to be located on the ground plane. The associated uncertainty is a superposition of the feature position uncertainty and the propagated prediction uncertainty[2]. We linearize the ground plane induced projection at the current estimate and apply linear error propagation [31]

$$\widehat{\boldsymbol{\Sigma}}_g = \boldsymbol{\Sigma} + \left(\left.\frac{\partial \mathbf{h}(\mathbf{u}, \boldsymbol{\xi})}{\partial \boldsymbol{\xi}}\right|_{\boldsymbol{\xi}=\widehat{\boldsymbol{\xi}}} \right) \mathbf{P} \left(\left.\frac{\partial \mathbf{h}(\mathbf{u}, \boldsymbol{\xi})}{\partial \boldsymbol{\xi}}\right|_{\boldsymbol{\xi}=\widehat{\boldsymbol{\xi}}} \right)^T. \tag{6.5}$$

Similarly, the predicted position $\widehat{\mathbf{u}}'_o$ and associated covariance matrix $\widehat{\boldsymbol{\Sigma}}_o$ of a point with height difference Δh compared to the ground can be computed using

[2] We assume that the errors in feature matching only corrupt the position of the corresponding feature in the successive image.

Figure 6.2: Planar parallax vectors superimposed on an image of a backward-facing camera. Feature points are shown in blue. The planar parallax with respect to the estimated ground plane is shown in orange. Notice the non-vanishing parallax on the left sidewalk. For better visualization, some feature mismatches have been removed. To visualize the epipolar field property, epipolar lines have been superimposed in red. Predicted feature positions above the horizon are caused by plane-ray intersections corresponding to antipodal rays.

equation (6.2). The predicted positions and uncertainties are illustrated in Figure 6.3. The most challenging structural outliers occurring in automotive applications are planes with similar parameters as the ground plane such as sidewalks. Thus, we assume that $\widehat{\mathbf{u}}'_o$ is located on a slightly elevated virtual parallel plane (a virtual sidewalk). A pair of image correspondences is then identified as an inlier if it complies with the prediction $\widehat{\mathbf{u}}'_g$ on the one hand, and is unlikely to corre-

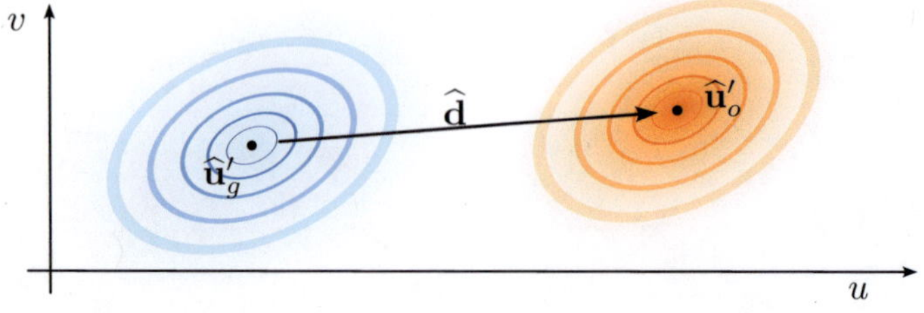

Figure 6.3: Prediction of single feature position under ground plane hypothesis, $\widehat{\mathbf{u}}'_g$ (blue), and object hypothesis, $\widehat{\mathbf{u}}'_o$ (orange, cf. Figure 6.1). The parallax vector is given by $\mathbf{d}$. Ellipsoids indicate isocontours of the corresponding probability density functions.

spond to a 3D point on the virtual parallel plane, on the other hand. A feature point $\mathbf{u}'$ in the successive image is then accepted as inlier if it is in the set

$$\mathcal{M}_\rho = \left\{ \mathbf{u}' \mid \left(\mathbf{u}' - \widehat{\mathbf{u}}'_g\right)^T \widehat{\boldsymbol{\Sigma}}_g^{-1} \left(\mathbf{u}' - \widehat{\mathbf{u}}'_g\right) \leq \rho \right\}, \tag{6.6}$$

i.e. the set of points for which the squared Mahalanobis distance between $\mathbf{u}'$ and $\widehat{\mathbf{u}}'_g$ is smaller than or equal to a threshold ρ. The expected false positive rate ν with regard to $\widehat{\mathbf{u}}'_o$ on the virtual parallel plane is given by the integral of the Gaussian density function $p\left(\mathbf{p}; \widehat{\mathbf{u}}'_o, \widehat{\boldsymbol{\Sigma}}_o\right)$ over the set of accepted correspondences (positives)

$$\nu := \int_{\mathcal{M}_\rho} p\left(\mathbf{p}; \widehat{\mathbf{u}}'_o, \widehat{\boldsymbol{\Sigma}}_o\right) d\mathbf{p}. \tag{6.7}$$

Here, $p\left(\mathbf{p}; \widehat{\mathbf{u}}'_o, \widehat{\boldsymbol{\Sigma}}_o\right)$ is the probability density at $\mathbf{p}$. Note that the expected false positive rate defined here only refers to the specific case where samples drawn from the distribution corresponding to $\mathbf{u}'$ fall inside the acceptance region. Equation (6.7) yields an implicit definition of ρ.

In order to derive an explicit expression we make simplifying assumptions. For small parallax vectors it is reasonable to assume that the covariance matrices $\widehat{\boldsymbol{\Sigma}}_g$ and $\widehat{\boldsymbol{\Sigma}}_o$ do not differ significantly. Thus we can replace $\widehat{\boldsymbol{\Sigma}}_g$ and $\widehat{\boldsymbol{\Sigma}}_o$ by $\widehat{\boldsymbol{\Sigma}}$. The implicit definition of ρ in equation (6.7) can then be transformed into an explicit one. To this end, we apply an affine transformation which maps $p\left(\mathbf{p}; \widehat{\mathbf{u}}'_g, \widehat{\boldsymbol{\Sigma}}\right) \mapsto p\left(\mathbf{p}; \mathbf{0}, \mathbf{I}_{2\times 2}\right)$ and $\widehat{\mathbf{u}}'_o \mapsto (\gamma, 0)^T$, i.e. a point on the u-axis. The corresponding transformed random process is given by the two-vector $(U, V)^T$ of independent standard normal distributed random variables with mean $(\gamma, 0)^T$.

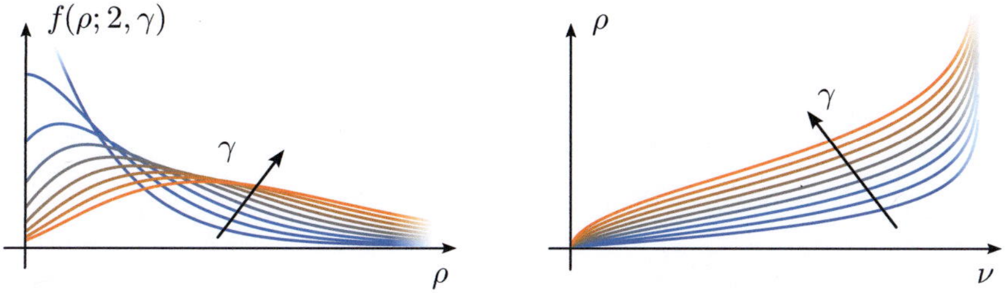

Figure 6.4: Graph of the probability density function of the noncentral χ^2 distribution (left) and inverse cumulative noncentral χ^2 distribution (right) for increasing noncentrality coefficients.

The sum of squares is distributed according to the noncentral χ^2 distribution with two degrees of freedom [2]

$$U^2 + V^2 \sim f(\rho; 2, \gamma), \tag{6.8}$$

and noncentrality coefficient

$$\gamma = \widehat{\mathbf{d}}^T \, \widehat{\boldsymbol{\Sigma}}^{-1} \widehat{\mathbf{d}}. \tag{6.9}$$

Note that the noncentrality coefficient coincides with the u-value of the transformation of $\widehat{\mathbf{u}}'_o$. The graph of the corresponding probability density function $f(\rho; 2, \gamma)$ is shown on the left side of Figure 6.4 for different noncentrality coefficients. The threshold is then computed from the inverse cumulative noncentral χ^2 distribution

$$\rho = F^{-1}(\nu; 2, \gamma). \tag{6.10}$$

The graph of the inverse cumulative noncentral χ^2 distribution for different noncentrality coefficients is shown on the right side of Figure 6.4. In summary, to determine the threshold for an image point $\mathbf{u}$, the positions and uncertainties in the successive view for a corresponding 3D point on the ground plane and on the virtual plane are predicted. From this, the noncentrality coefficient (equation (6.9)) is computed and finally the inverse cumulative noncentral χ^2 distribution is evaluated. Figure 6.5 shows a threshold image for two cameras mounted on a vehicle facing to the right and backward, respectively. Motion and ground plane estimates are provided by a Kalman filter. It can be seen that high and low thresholds form regions with smooth transitions between them. Around the

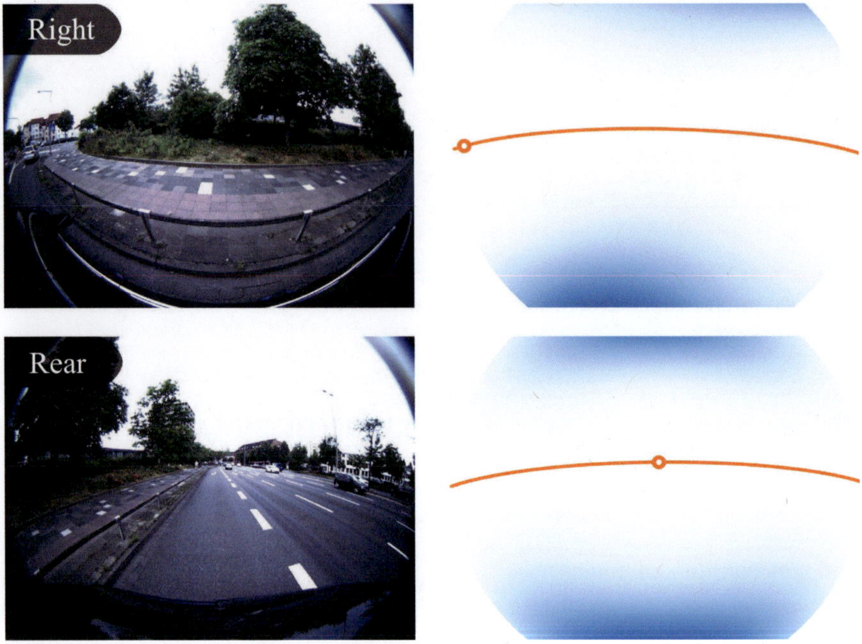

Figure 6.5: Two example images and corresponding threshold images for a camera facing to the right (top) and backward (bottom). Blue color intensity indicates threshold values. The horizon and epipole are depicted by an orange line and point, respectively. See text for details on the color scheme.

epipoles and the horizon the thresholds are small. Non-vanishing thresholds above the horizon are caused by plane-ray intersections of antipodal rays and can be disregarded. A major drawback of the approach presented so far is that it yields non-vanishing thresholds for vanishing parallax. In case that the predicted parallax is zero, $\widehat{\mathbf{d}} = \mathbf{0}$, the predicted image positions $\widehat{\mathbf{u}}'_g$ and $\widehat{\mathbf{u}}'_o$ as well as the associated probability density distributions coincide (cf. Figure 6.3). For any non-zero false positive rate, $\nu \neq 0$, equation (6.10) then yields a non-zero threshold value. This implies the probability of accepting an outlier is equal to the probability of accepting an inlier in the special case of zero parallax. This undesired property can be avoided by neglecting feature correspondences with small predicted parallax.

6.3 Sequential Testing and Updating

In the previous section local adaptive thresholds have been derived. In the following we show how presorting feature point correspondences by their associated threshold values in combination with a sequential processing scheme improve the robustness of our approach.
The predicted true positive rate η, i.e. the probability of correctly identifying an inlier, can be computed from the threshold ρ as

$$\eta = F(\rho; 2, 0), \tag{6.11}$$

i.e. the noncentral χ^2 distribution with $\gamma = 0$ (centered χ^2 distribution). As $F(\cdot)$ is monotonically increasing, feature point correspondences with associated high threshold are more likely to be correctly identified as inliers. This property can be exploited by first testing point correspondences with high thresholds and, if these are identified as inliers, incorporate them into the state estimation prior to testing correspondences with lower thresholds. Correspondences with lower threshold are then tested based on consolidated estimates. This approach can easily be embedded into a Kalman filter by applying the sequential updating scheme [6]. The sequential processing steps are then as follows.
First, we compute threshold values for the whole set of N feature point correspondences based on the a priori state estimate, yielding 3-tuples $(\mathbf{u}_i, \mathbf{u}'_i, \rho_i)$, with $i = 1, ..., N$. The tuples are then ordered by their threshold values and the one with highest threshold is taken from the set and tested. If an inlier is found, it is used to update the state. A consolidated state estimate is available once an inlier is found. Therefore, threshold values are recomputed prior to testing. This approach mitigates the risk of rejecting inliers, as more sensitive data is tested based on consolidated estimates.
To evaluate the robustness of our approach, we have recorded a sequence in the inner city that contains typical structural outliers, such as sidewalks. The camera was attached to the right side of the vehicle. While driving, the vehicle laterally approaches a sidewalk. Feature point correspondences between successive views were established and used as input to an Extended Kalman Filter with local adaptive thresholds and sequential processing scheme. The Kalman filter was initialized using rough estimates of the motion and ground plane, and the false positive rate ν and height difference Δh were set to 5% and -75mm, respectively. Inliers detected by our approach are shown in blue superimposed on the images in the right column in Figure 6.6. The number of identified inliers is given in the top right side, respectively.

For comparison, we applied a RANSAC-based approach [31] for homography estimation to the same set of feature correspondences. The global threshold of the RANSAC-based approach was chosen such that the number of incorrectly identified inliers on the sidewalk in frame 633 is about the same as for our approach. As RANSAC is designed to find the largest consensus set, the ground plane can only be detected if it is the dominant structure in the scene. The largest consensus set found by RANSAC is shown in orange superimposed on the images in the right column in Figure 6.6. It can be seen that most inliers are found on the sidewalk in frame 648, and on a virtual plane in frame 765. After removing the largest consensus set RANSAC has been reapplied to the remaining feature points. The second largest consensus set is shown in green for frame 648, and the third largest consensus set is shown in green for frame 765. The second largest consensus set in frame 765 corresponded to a virtual plane and is not shown here. From the inlier count we can see that the RANSAC-based approach with global threshold detects significantly less inliers and fails in the complex scenario in frame 765. Note that the robust sequential processing approach with local adaptive threshold performs significantly better than the basic RANSAC-based approach in the considered scenario.

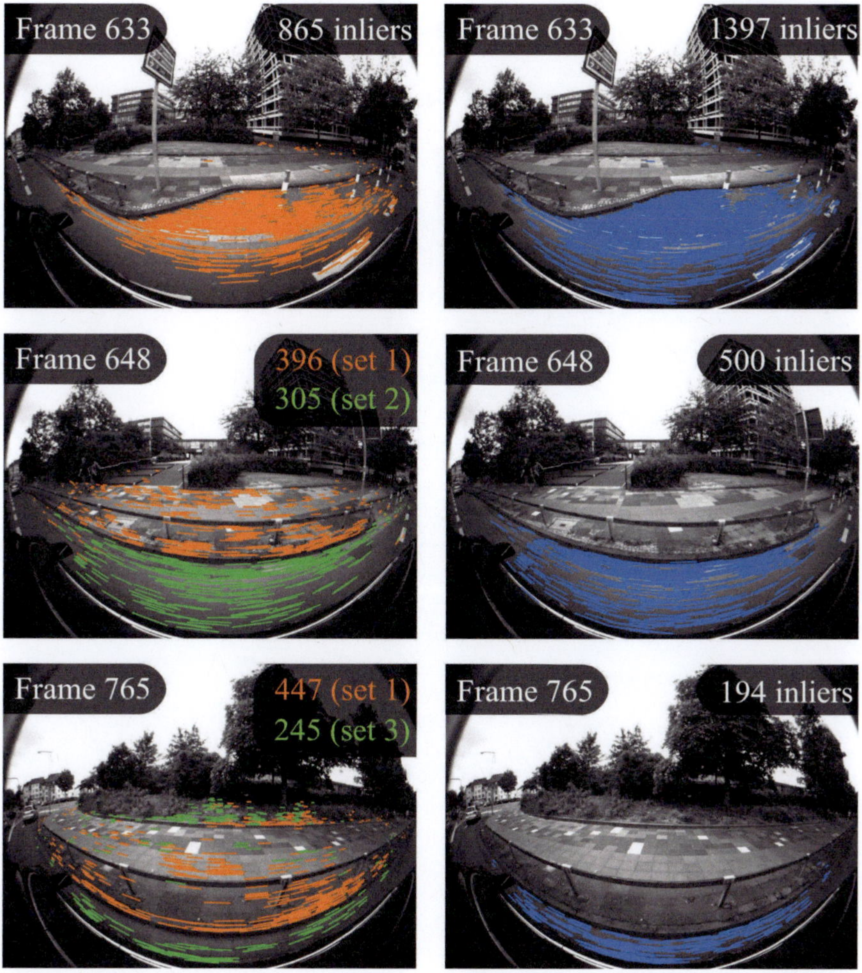

Figure 6.6: Results of RANSAC-based homography estimation approach (left column) and the robust sequential processing approach (right column). The threshold for RANSAC is chosen such that the number of incorrectly identified inliers on the sidewalk in frame 633 is about the same in both cases. See text for color scheme. This figure includes images taken from [45].

7 Continuous Self-Calibration Based on Kalman Filtering

In the previous chapters we have built the theoretic foundation for extrinsic camera calibration and presented required key elements. In this chapter we combine these results and present a novel algorithm for continuous extrinsic self-calibration. We seek to estimate the extrinsic calibration parameters by combining the constraints arising from relative camera motions, the epipolar geometry of rigidly coupled cameras, and the observation of the ground plane in multiple views. This problem can be approached in different ways. A short overview with regard to filtering and general optimization techniques was given in Section 2.4. A further distinction can be made with respect to the problem formulation. Esquivel et al. [19] and Pagel et al. [76, 75], for instance, first estimate the motion of each individual camera explicitly and then apply hand-eye calibration to obtain estimates of the extrinsic calibration. These estimates are treated as measurements in subsequent processing. Other approaches ([18, 70, 30, 103, 69]) use an implicit formulation, i.e. the parameters and observations are subject to implicit measurement constraints. The implicit formulation has several advantages. Measurement constraints arising from other sensors can be incorporated easily. The implicit formulation does not require each camera to be able to estimate its motion, and problem of scale drift, inherent to monocular systems, can typically be mitigated. The latter two properties make this formulation typically more robust. However, a disadvantage of the the implicit formulation is that it requires an initial estimate. Because of the aforementioned advantages we use an implicit formulation.

Kalman filters are well-suited for this type of problem and have been used successfully for similar calibration problems in the past [18, 30, 75, 69]. The filtering property allows processing new data as it arrives, thus enabling continuous (online) processing.

In contrast to other approaches we do not track feature correspondences over multiple frames. Instead we only use frame-to-frame image point correspondences and avoid computation of respective 3D structure. As a result, the size of the state vector is small compared to classical structure from motion methods.

A motion model is required for the temporal update of the Kalman filter. In Chapter 4 we have seen that the extrinsic calibration parameters can neither be estimated from straight motions nor from circular motions. Typically, the motion of road vehicle is mostly planar, rendering the estimation of all extrinsic calibration parameters difficult if we do not incorporate the constraints imposed by the ground plane. For this reason, we restrict ourselves to the case where all cameras are able to observe the ground plane. In this case, both, planar motion and general motion allow estimating all parameters.

We first give a brief overview of the proposed algorithm before going into more detail. An extensive evaluation of the self-calibration algorithm is presented in Chapter 8. A preliminary work of the extrinsic self-calibration algorithm presented here has been published in [44].

7.1 Recursive Filtering

In this section we introduce the extended Kalman filter equations and give a brief overview of our recursive filtering approach. We apply a single extended Kalman filter [6]. The motion and ground plane parameters, as well as the extrinsic calibration parameters are associated with a state vector $\boldsymbol{\xi}$ of a dynamic system which evolves, corresponding to a discrete time nonlinear stochastic system

$$\boldsymbol{\xi}_k = \mathbf{f}\left(\boldsymbol{\xi}_{k-1}\right) + \mathbf{q}_k, \tag{7.1}$$

where $\mathbf{q}_k$ denotes the process noise, which we assume to be zero mean and Gaussian $\mathbf{q}_k \sim \mathcal{N}\left(\mathbf{0}, \mathbf{Q}_k\right)$. In addition, we assume the measurements to be perturbed by additive zero mean Gaussian noise,

$$\mathbf{z}_k = \overline{\mathbf{z}}_k + \mathbf{w}_k, \tag{7.2}$$

where $\overline{\mathbf{z}}_k$ is the error free measurement vector and $\mathbf{w}_k \sim \mathcal{N}\left(\mathbf{0}, \mathbf{W}_k\right)$. We further assume $\mathbf{q}_k$ and $\mathbf{w}_k$ to be mutually uncorrelated. We use the more general, implicit formulation of the measurement functions.

The error free observations satisfy

$$\mathbf{m}\left(\boldsymbol{\xi}_k, \overline{\mathbf{z}}_k\right) = \mathbf{0}, \tag{7.3}$$

where $\mathbf{m}\left(\cdot,\cdot\right)$ are nonlinear measurement constraint equations which will be introduced in Section 7.3. Since both, the state transition function (equation (7.1)) and the measurement constraints (equation (7.3)) are nonlinear, an extended Kalman filter instead of a (linear) Kalman filter is applied. The a priori and a posteriori state estimates are given by $\widehat{\xi}_k^-$ and $\widehat{\xi}_k^+$, respectively, and the associated covariance matrices are given by $\mathbf{P}_k^-$ and $\mathbf{P}_k^+$, respectively. The complete set of extended Kalman filter equations can be found in Appendix A.5.

In the following we give an overview of the extended Kalman filter self-calibration algorithm. To this end, we consider the flowchart in Figure 7.1. The figure illustrates one cycle of the extended Kalman filter. In the center column the evolution of the state vector and associated covariance matrix is shown. In the left and right column the processing steps for overlapping fields of view and motion and ground plane estimation are shown, respectively.

At the beginning of each cycle the state prediction and associated covariance matrix are computed. The state vector elements corresponding to the extrinsic calibration remain unchanged during this step. The a priori state estimate $\widehat{\xi}_k^-$ is then used to warp captured images with the objective to make corresponding image regions coincide (cf. Chapter 5). It should be noted that leveraging the a priori estimate for image prewarping is only reasonable if the estimate is closer to the true state than the assumption of identical camera coordinate frames, the latter being typically followed when conducting feature extraction and matching. Putative image correspondences are established using the FAST corner detector and the BRIEF feature descriptor. To detect inliers to the epipolar geometry we employ a classic Random Sample Consensus (RANSAC) algorithm [20]. An essential matrix is computed from five randomly drawn pairs of putative correspondences using the algorithm of Nistér [72]. The algorithm provides up to ten solutions which are tested against the whole set of putative correspondences. Inliers are selected based on a Sampson error criterion [31]. The largest set of inliers among the ten solution and multiple repetitions is then selected for further processing.

The independence of the inlier detection from the current filter state has proven to be advantageous, especially in the beginning when the error of the state estimate is still large. During the update stage we make intensive use of the sequential processing scheme [6]. Instead of updating the state vector using all measurements simultaneously, the measurements are processed sequentially. This allows to first incorporate the measurements from overlapping fields of view and motion estimation based on epipolar geometry before using the intermediate state estimate for robust homography estimation (cf. Chapter 6). Opposed to the approach presented in Chapter 6 we apply a decomposition of the measurement residual. We decompose the residual into two orthogonal parts, perpendicular to the epipolar line (epipolar distance) and along the epipolar line.

This step is required in order to obtain a partially updated state vector without incorporating the same measurement twice. We elaborate on the decomposition and the extension to epipolar curves in Section 7.3.

7.2 Parameterization and Motion Models

The findings of Chapter 4 indicate that planar as well as general motion combined with the ground plane as a reference object enable the estimation of all extrinsic calibration parameters. General motion is preferable in this case as it allows observing all parameters using only two consecutive poses of the multi-camera system. However, typically motion of road vehicles is mostly planar. An experimental comparison of both models is presented in Chapter 8.

The parameterization of the extended Kalman filter plays a fundamental role. It should be locally continuous and differentiable. Furthermore, the extended Kalman filter can become unstable if the assumption of local linearity is violated. In this context, the parameterization of rotation matrices and unit vectors constitutes a particular, but well-studied problem. A minimal parameterization of rotation matrices and unit vectors is desirable for two reasons, the first being the reduction of the state vector dimensionality, and the second being the avoidance of constraints in the state space that require special and careful treatment [39]. Unfortunately, all 3-vector parameterizations of rotation matrices and 2-vector parameterizations of a point on a three dimensional sphere contain singularities (e.g. the gimbal lock for rotation matrices). However, in the vicinity of the origin these parameterizations behave well and adhere to the above requirements.

In this thesis we assume that an initial state estimate is provided. We use the estimate to apply a normalization transform on the state vector. In consequence, the state vector only contains the deviations from the initial estimate. For example, to compute the relative orientation between a camera and the reference camera the orientation matrix corresponding to the current state vector is multiplied with thc respective denormalizing orientation matrix. After computing the normalization transform from the initial state estimate the respective elements in the state vector are set to zero. If the initial state estimate is sufficiently close to the ground truth, singular configurations are avoided. To represent rotation and orientation matrices we apply a minimal 3-vector parameterization. Without loss of generality, we use the Cayley transform [25]. The Cayley transform is closely related to quaternions[1] and has a singularity rotations through 180°. To represent 3D unit vectors we use spherical coordinates and apply either rotation matrices or Householder reflections [25] for normalization. A generalization of this approach is the multiplicative extended Kalman filter (MEKF) [60]. The MEKF updates the normalization transform at the end of each filter cycle. This is a standard procedure in current optimization frameworks (see e.g. [46]). However, we found the normalization with respect to the initial estimate to be sufficient. In the following we elaborate on the parameterization of relative camera poses and ground plane. Thereafter, we present the motion model specific parameterizations.

As presented in Chapter 4.1 we describe the extrinsic calibration via relative pose transformations between the cameras and the reference camera. The motion of the camera system is expressed in the coordinate frame of the reference camera. The relative orientation between the cameras are parameterized using the Cayley transform. The displacements are parameterized directly by 3-vectors. A normalization is not required in this case. To fix the scale of the multi-camera system the baseline between the reference camera and a dedicated second camera is kept constant. This baseline is not part of the state vector. The displacement direction of the dedicated camera is parameterized using spherical coordinates. The ground plane normal is parameterized in the same way and the camera height is represented by a scalar.

We employ a constant velocity model [6], i.e. any changes in the velocity are modeled by process noise. Since the system has no means to determine its position or orientation with respect to a world frame we do not track these parameters. Discrete planar motion can be described using three parameters, one for the translation direction and two for the rotational and translational velocity. We construct a ground plane coordinate frame by projecting the principal axis onto the ground plane (see Figure 7.2). The translation direction is then defined by the azimuth angle. During the state prediction the parameters of the planar motion model remain unchanged.

General motion has six degrees of freedom. Similar to the planar motion model, we make use of a ground plane aligned coordinate frame in which the translation and rotation are described. In contrast to the planar motion model the height and ground plane normal are adjusted according to the out-of-plane translation and rotation during the state prediction. Table 7.1 summarizes the distribution of parameters for both models.

[1] The real element is set to one.

	Number of parameters	
	General motion	Planar motion
Rotation	3	1
Translation	3	2
Ground plane	3	3
Relative pose transformations	6 (C-1)	6 (C-1)
Scale fixing	-1	-1
State dimension for $C = 4$	26	23

Table 7.1: Distribution of the state vector elements for both motion models with respect to the number of cameras C.

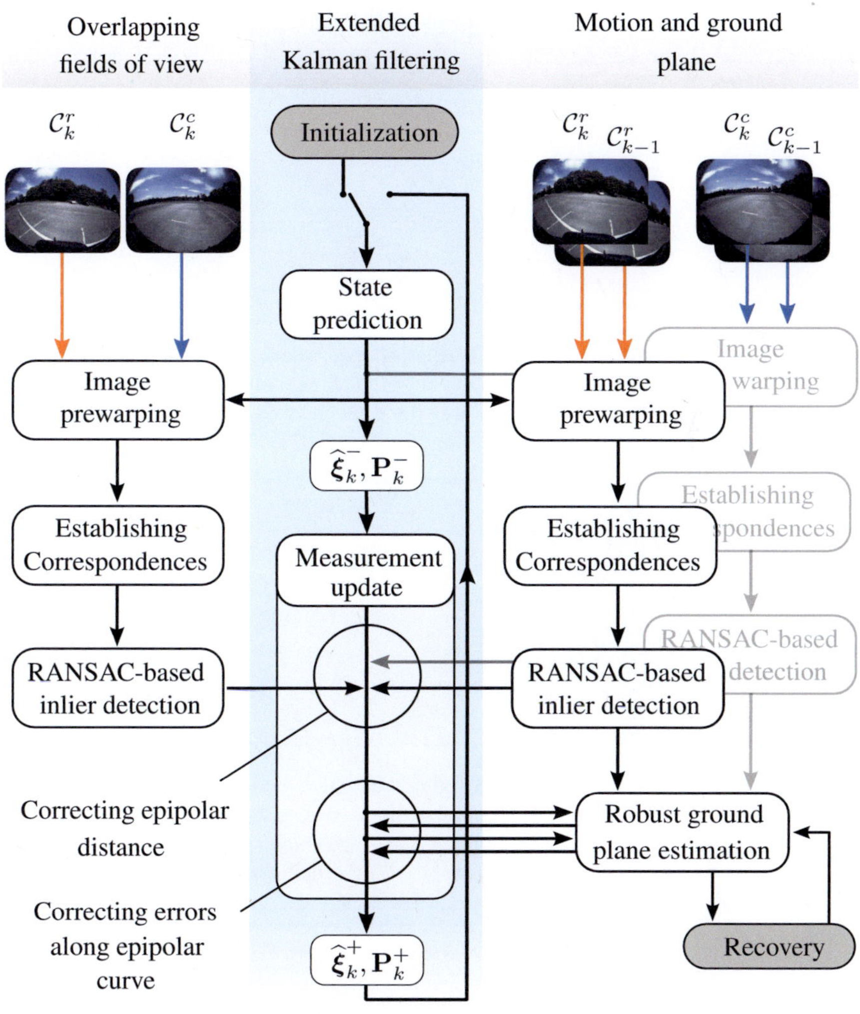

Figure 7.1: Flowchart illustrating one cycle of the extended Kalman filter. The evolution of the state estimate and associated covariance matrix is shown in the center column. To the left and right the flow chart for overlapping fields of view and motion and ground plane estimation are shown, respectively. Initialization and recovery (highlighted in gray) are executed only when necessary.

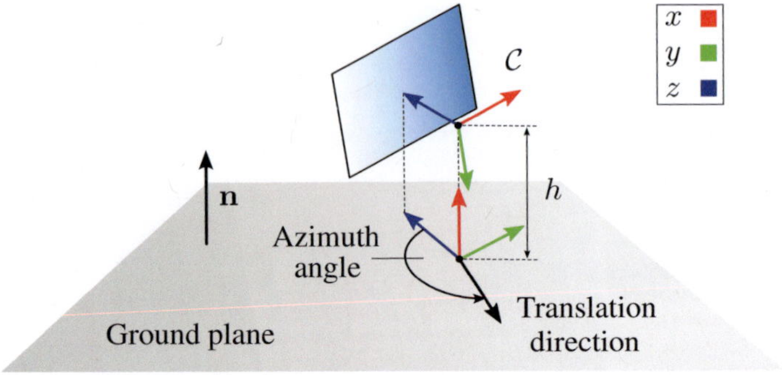

Figure 7.2: Definition of the ground plane coordinate frame. The x-axis of the ground plane coordinate frame is aligned with the current estimate of the ground plane normal, and the z-axis aligned with the projection of the principal axis onto the ground plane.

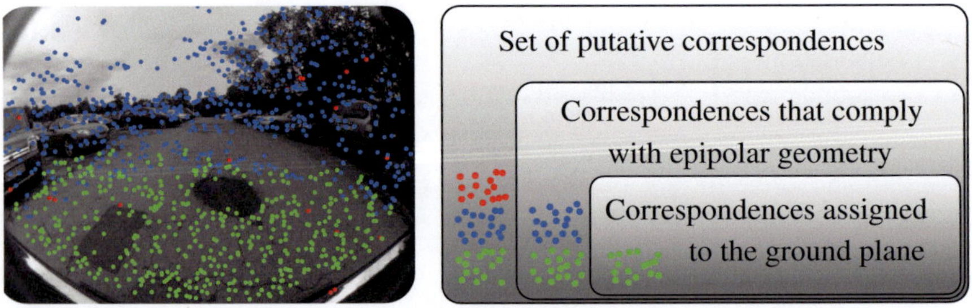

Figure 7.3: Exemplary output of interest points for which a putative correspondence was found in the next image (left). The set of all putative correspondences can be divided into correspondences that comply with epipolar geometry (blue, green) and outliers which do not (red). Correspondences that comply with the estimated epipolar geometry can further be assigned to the set of correspondences that are associated with the ground plane (green).

7.3 Extended Kalman Filter Update Stage

During the update stage of the extended Kalman filter a sequential updating scheme is employed, i.e. assuming the measurement noise to be uncorrelated (i.e. the covariance matrix has block diagonal structure), an update can be performed for each measurement individually [6]. One advantage of this approach is that the inversion of large matrices can be avoided. More importantly, this allows applying a stratified approach in which the information of a partially updated state vector is used during inlier detection in a subsequent algorithm stage. The sequential processing algorithm is presented in Appendix A.6.
The set of all putative image correspondences consists of those that comply with the epipolar geometry, with the ground plane homography, or are treated as outliers. All correspondences that are associated with the ground plane also satisfy the epipolar constraints. This is illustrated in Figure 7.3. We first update the state estimate and covariance matrix using correspondences that comply with the epipolar geometry. Since correspondences have been selected using a robust method that is independent of the current state estimate this stage does not benefit from a partially updated state vector. However, robust homography estimation does. In a second step, the partially updated state vector and preselected correspondences are used to update the ground plane estimate (cf. Figure 7.1) which also adopts the sequential updating scheme.
The algorithm presented in Chapter 6 estimates the camera motion as well as the ground plane and imposes two constraints on the feature point positions. Since estimating the epipolar geometry already imposes one constraint on the feature point positions we have to modify the algorithm presented in Chapter 6 in order to avoid incorporating the same measurement twice. To this end, we apply a decomposition of the measurement residual.
Let the point correspondence $\mathbf{u} \leftrightarrow \mathbf{u}'$ with associated covariance matrix $\boldsymbol{\Sigma}$ and an estimate of the epipolar geometry and the ground plane be given. We apply a linear approximation of the epipolar curve at a support point

$$\mathbf{u}'_s = \boldsymbol{\kappa}\left(-\widehat{\mathbf{E}}\mathbf{x} \times \left(\widehat{\mathbf{E}}\mathbf{x} \times \mathbf{x}'\right)\right), \tag{7.4}$$

i.e. the image of the ray $\mathbf{x}'$ projected onto the epipolar plane defined by the corresponding ray $\mathbf{x}$ in the first view and the estimated essential matrix. Recall that the rays corresponding to $\mathbf{u}$ and $\mathbf{u}'$ are $\mathbf{x}$ and $\mathbf{x}'$, respectively. The line $\widehat{\mathbf{l}}'$ is the tangent of the epipolar curve in $\mathbf{u}'_s$. This is depicted in Figure 7.4. In general $\mathbf{u}'_s$ is not the closest point on the epipolar curve (this is emphasized

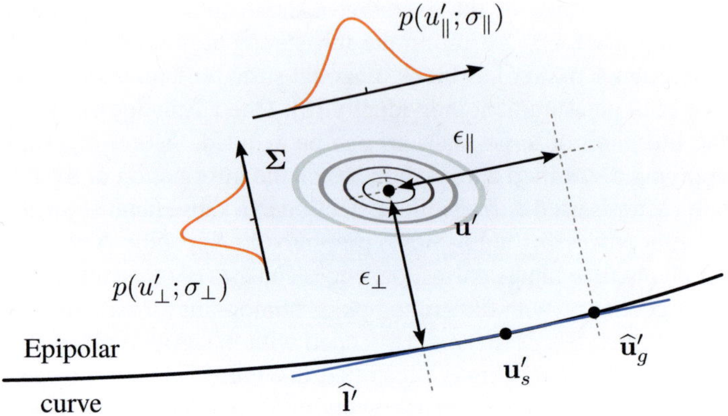

Figure 7.4: Decomposition of the measurement residual. The epipolar curve is linearized at a support point $\mathbf{u}'_s$ yielding $\widehat{\mathbf{l}}'$. The approximated geometric image distance $\epsilon_{\perp}$ is the distance between $\mathbf{u}'$ and $\widehat{\mathbf{l}}'$. The point $\widehat{\mathbf{u}}'_g$ is the ground plane induced prediction. The approximated error along the epipolar curve is given by $\epsilon_{\|}$. The marginalization of the covariance matrix $\boldsymbol{\Sigma}$ is carried out with respect to $\widehat{\mathbf{l}}'$ and its normal. The marginal distributions are $p(u'_{\|};\sigma_{\|})$ and $p(u'_{\perp};\sigma_{\perp})$.

in Figure 7.4). Using the direction of $\widehat{\mathbf{l}}'$ and its normal, we marginalize the covariance matrix, yielding $p(u'_{\perp};\sigma_{\perp})$ and $p(u'_{\|};\sigma_{\|})$. If the associated 3D point is located on the ground plane we compute the prediction $\mathbf{u}'_g$ which is located on the estimated epipolar curve.
We define the measurement constraints functions (7.3) as

$$m_{\perp}\left(\widehat{\boldsymbol{\xi}},\mathbf{u}'\right)=\frac{\widehat{l}'_1 u'+\widehat{l}'_2 v'+\widehat{l}'_3}{\sqrt{(\widehat{l}'_1)^2+(\widehat{l}'_2)^2}}=\epsilon_{\perp} \tag{7.5}$$

and

$$m_{\|}\left(\widehat{\boldsymbol{\xi}},\mathbf{u}'\right)=\frac{\widehat{l}'_2(u'-\widehat{u}'_g)-\widehat{l}'_1(v'-\widehat{v}'_g)}{\sqrt{(\widehat{l}'_1)^2+(\widehat{l}'_2)^2}}=\epsilon_{\|}, \tag{7.6}$$

respectively, where $\widehat{\mathbf{l}}' = (\widehat{l}'_1, \widehat{l}'_2, \widehat{l}'_3)^T$. Equation (7.5) approximates the epipolar distance and equation (7.6) approximates the error along the epipolar curve (cf. Figure 7.1).

7.4 Initialization and Recovery of Vehicle Velocity

The ratio of vehicle velocity and height has direct influence on the distance of the feature point displacement in the image. Since the initially provided camera height and vehicle velocity may deviate significantly from the actual values (e.g. by up to one order of magnitude), the robust homography estimation algorithm presented in Chapter 6 may not be able to detect and track the ground plane, causing the calibration algorithm to diverge.

In the following we present a method for initialization and recovering the vehicle velocity to height ratio. We assume the translation direction, camera rotation, and the ground plane to be known sufficiently well. These conditions are usually fulfilled as long as at least one camera is able to track the ground plane. The output of the method is an estimate of the velocity to height ratio which can be used to detect ground plane inliers.

Let the rotation matrix $\mathbf{R}$, the translation direction $\mathbf{t}/\|\mathbf{t}\|_2$, and the ground plane normal $\mathbf{n}$ be known. We are searching for the velocity to height ratio $\tau = \|\mathbf{t}\|_2/h$. We formulate the search as a least square problem

$$\widehat{\tau} = \arg\min_{\tau} \left\{ \Bigg(\underbrace{\mathbf{X}' - \mathbf{R}\mathbf{X} + \mathbf{n}^T\mathbf{X}\frac{\mathbf{t}}{\|\mathbf{t}\|_2}\tau}_{\mathbf{v}(\tau)} \Bigg)^2 \right\}, \tag{7.7}$$

where $\mathbf{X}$ and $\mathbf{X}'$ are corresponding 3D measurements of the same point on the ground plane of a moving camera. The estimate $\widehat{\tau}$ minimizes the squared Euclidean distance between the 3D points.

Since the 3D points $\mathbf{X}'$ and $\mathbf{X}$ are in general not known but assumed to be located on the ground plane, we intersect the corresponding rays and ground plane using equation (3.13). The plane normal in the second view can be computed using equation (4.5). Finally, we make use of equation (4.7) to determine the

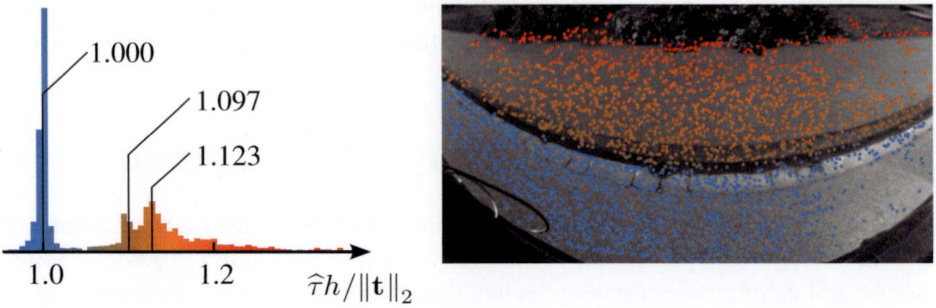

Figure 7.5: Estimation of velocity to height ratio τ. Each pair of corresponding feature points contributes to one estimate. The histogram of estimates is shown on the left. For better readability the x-axis has been scaled such that estimates of the correct ratio accumulate around one. Using a mean shift algorithm, three modes have been detected and the corresponding values are shown. The corresponding image with colored interest points is shown on the right hand side. The same color coding has been applied. A clear distinction between the ground plane and the sidewalk can be made, in both, the histogram and the image.

height ratio of the cameras center in successive frames. After substitution and reorganization we obtain

$$\mathbf{v}_n(\tau) = \underbrace{\frac{\mathbf{x}'}{\mathbf{n}^T\mathbf{R}^T\mathbf{x}'} - \mathbf{R}\frac{\mathbf{x}}{\mathbf{n}^T\mathbf{x}}}_{\mathbf{v}_0} + \underbrace{\left(\mathbf{I}_{3\times 3} - \frac{\mathbf{x}'}{\mathbf{n}^T\mathbf{R}^T\mathbf{x}'}\mathbf{n}^T\mathbf{R}^T\right)\frac{\mathbf{t}}{\|\mathbf{t}\|_2}}_{\mathbf{v}_\tau}\tau, \quad (7.8)$$

where $\mathbf{v}_n(\tau) = -\mathbf{v}/h$ is normalized by the height. Note that $\mathbf{v}(\tau)$ (equation (7.7)) and $\mathbf{v}_n(\tau)$ (equation (7.8)) yield the same estimate $\widehat{\tau}$. The complete derivation of equation (7.8) can be found in Appendix A.7. The least squares solution is then given by

$$\widehat{\tau} = -\mathbf{v}_0^T\mathbf{v}_\tau/\mathbf{v}_\tau^T\mathbf{v}_\tau. \quad (7.9)$$

An estimate $\widehat{\tau}_i$ can be computed for each pair of corresponding image points $\mathbf{u}_i \leftrightarrow \mathbf{u}'_i$, where $i = 1...N$. We interpret the estimates as samples of a probability distribution and apply a mean shift algorithm [17] to detect the mode. An example for a well calibrated multi-camera system is shown in Figure 7.5. We execute this algorithm if the number of correspondences that comply with the current estimate the ground plane induced homography is below a heuristically chosen threshold. The most common reason for this is that the estimate of the

camera height is not consistent with the other cameras in the setup. We use a heuristic to decide whether a dominant mode was found. If this is the case we apply robust variance estimation by means of median absolute deviation to estimate the variance. The estimate and variance are then used as the basis for our robust homography estimation algorithm.
During initialization a modified version of this algorithm is executed. To adjust the initially provided velocity, the velocity to height ratio estimates from all cameras are normalized with respect to the corresponding camera heights and then combined to one set. The initial velocity is then set to the median value of the set. In the next chapter we show that this approach works well, even if the initial estimate of the ground plane normal is inaccurate.

8 Experimental Evaluation

In this chapter we present an extensive experimental evaluation of our extrinsic self-calibration algorithm. The evaluation is based on real-world data that was captured using a vehicle-mounted multi-camera system. Ground truth calibration parameters have been acquired by means of an offline calibration method. The ground truth serves as a reference to assess the self-calibration results quantitatively and to allow the comparison between different motion models, algorithm settings, and information sources.
In the following sections we introduce the evaluation dataset in detail and explain how ground truth was acquired, how the best parameter settings were found, and how the algorithm initialization was carried out. Thereafter, we present the quantitative evaluation and discuss approaches to assess the accuracy of the calibration at runtime. Finally, we show some qualitative results using three typical applications for multi-camera systems, namely visual odometry, generation of a virtual top view of the vehicle surrounding, and stereo rectification.

8.1 Evaluation Dataset

Our dataset consists of 24 sequences that have been recorded using four cameras that were mounted on a standard station wagon. The cameras were facing forward, to the left, backwards, and to the right. All cameras were equipped with identical fisheye lenses. Figure 8.1 illustrates the camera setup and respective fields of view. We employed standard industrial cameras with global shutter and 1.25 megapixels (1292×964 pixels). The cameras were synchronized to capture images simultaneously at 30Hz. The high recording frame rate allows to evaluate the performance of our algorithm at different frame rates by subsampling the image sequences. The horizontal angle of view of the camera-lens combination is approximately $185°$, resulting in large overlapping fields of view (see Figure 8.1). Similar setups have been used in [82, 36].
Figure 8.2 depicts the camera mounting positions and shows camera heights and relative distances. The smallest baseline between adjacent cameras is 2.3m (front to side), and the largest is 2.85m (rear to side). In contrast, the average camera height is less than half of the baselines. The mounting positions were

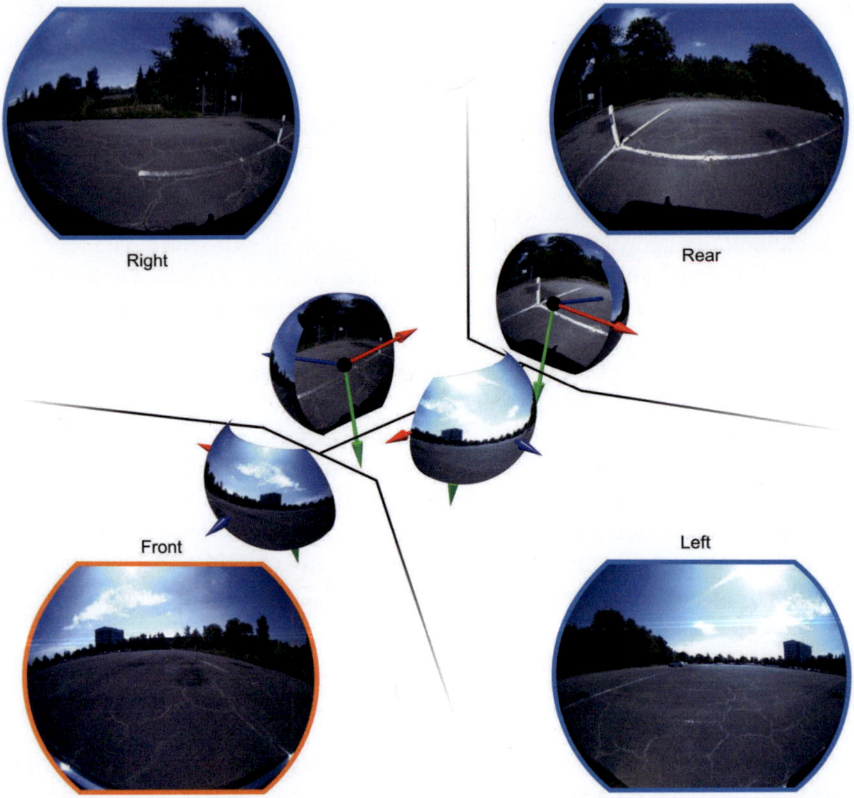

Figure 8.1: Illustration of the camera setup. The four simultaneously captured images are warped onto image spheres (center). The reference camera is marked orange. Adjacent cameras have overlapping fields of view. For example, the reflector post can be seen the upper two images despite being close to the vehicle.

chosen to resemble those of commercially available vehicles with multi-camera systems.

The 24 sequences of the dataset were recorded on one day during daytime in different parking areas. Parking areas were chosen as they represent a typical environment in which self-calibration function would be active. For example, the multi-camera system should be recalibrated after the vehicle is picked up or parked after production. Additionally, many of today's driver assistance systems

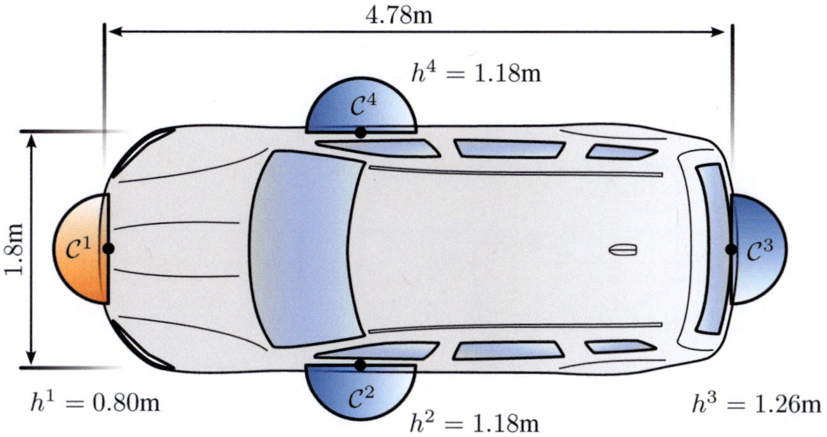

Figure 8.2: Illustration of camera mounting positions on the test vehicle, a standard station wagon. The reference camera is mounted in front. The left and right facing cameras (C^2 and C^4) are mounted close to the side mirrors. The distance between these cameras is approximately 1.8m and the distance between the front and rear-mounted camera is approximately 4.78m. The specified heights, h^1 to h^4, were determined during the (offline) reference calibration.

are designed to assist during the parking maneuver or to perform the task automatically. The driven trajectories resemble typical parking area maneuvers by containing e.g. low velocities, tight turns and nearby as well as distant objects. Figure 8.3 shows a subset of the driven trajectories. From the 24 sequences a subset of four was used for parameter tuning (shown in orange). The remainder was used for evaluation. The sequences contain between 723 and 2586 images per camera, corresponding to 24 to 86 seconds of recording-time. The shortest and longest track lengths are 112m and 568m, respectively. The total track length is around 5.8km and the average velocity is 19.4km/h, corresponding to an average of 0.18m per frame (at 30Hz).

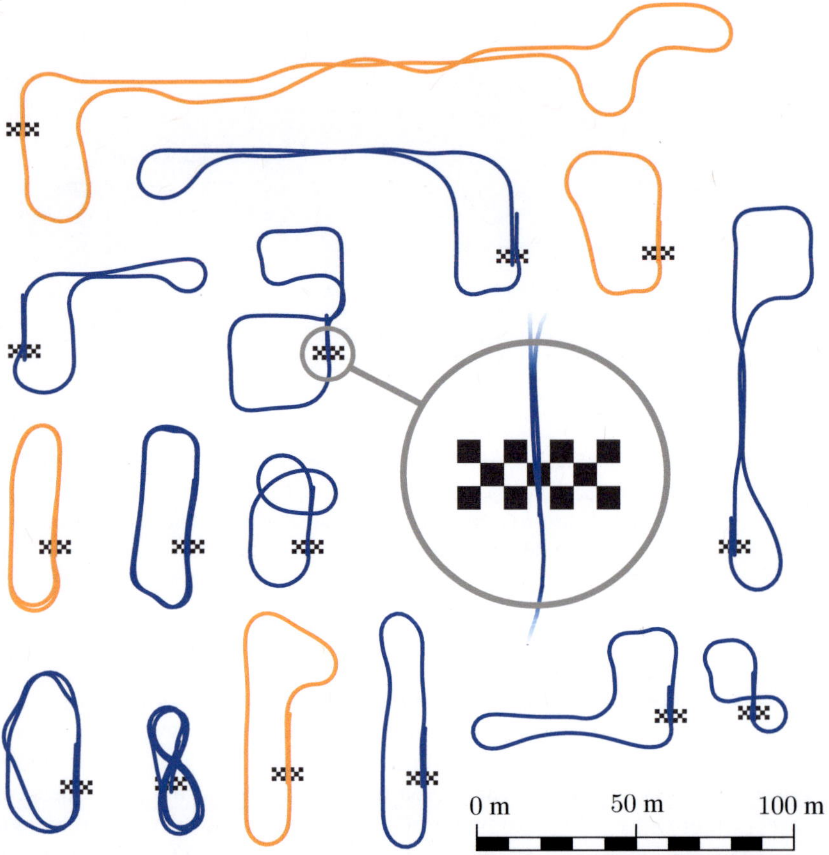

Figure 8.3: Estimated vehicle trajectories of recorded sequences. Visual odometry (cf. Section 8.5.1) was used to estimate the vehicle motion. A subset of the 24 sequences is shown here. The set of sequences is divided into a subset of four manually sequences for parameter tuning (orange) and the remainder of 20 sequences for evaluation (blue). The checkered flags mark the start of each recording and were passed at least twice during each recording.

8.2 Ground Truth and Error Metric

To evaluate the calibration results quantitatively a proper error metric and ground truth are required. In Section 4.1 we introduced the datum definition applied throughout this thesis. The camera coordinate frame of the front camera is chosen as the reference coordinate frame and the distance between the front and backward-facing cameras is chosen for scale fixing. To make the error metric independent of the choice of the reference coordinate frame and parameterization, we consider the estimated relative pose transformations between all cameras instead of considering only the transformations which have been estimated explicitly by the extended Kalman Filter.
Given the ground truth and estimated relative pose transformations between two cameras c and d,

$$\mathbf{\Delta T}_{gt}^{d\to c} = \begin{bmatrix} \mathbf{\Delta R}_{gt}^{d\to c} & \mathbf{\Delta t}_{gt}^{d\to c} \\ \mathbf{0}_{3\times 1}^{T} & 1 \end{bmatrix}, \mathbf{\Delta}\widehat{\mathbf{T}}^{c\to d} = \begin{bmatrix} \mathbf{\Delta}\widehat{\mathbf{R}}^{c\to d} & \mathbf{\Delta}\widehat{\mathbf{t}}^{c\to d} \\ \mathbf{0}_{3\times 1}^{T} & 1 \end{bmatrix}, \tag{8.1}$$

we compute the residual orientation angle

$$\epsilon_{\mathbf{\Delta R}}(c,d) = \cos^{-1}\left(\frac{\mathrm{tr}\left(\mathbf{\Delta R}_{gt}^{d\to c}\mathbf{\Delta}\widehat{\mathbf{R}}^{c\to d}\right) - 1}{2} \right) \tag{8.2}$$

and residual displacement length

$$\epsilon_{\mathbf{\Delta t}}(c,d) = \|\mathbf{\Delta R}_{gt}^{d\to c}\mathbf{\Delta}\widehat{\mathbf{t}}^{c\to d} + \mathbf{\Delta t}_{gt}^{d\to c}\|_2 \tag{8.3}$$

from the residual pose transformation $\mathbf{\Delta T}_{gt}^{d\to c}\mathbf{\Delta}\widehat{\mathbf{T}}^{c\to d}$, where $\mathrm{tr}(\cdot)$ denotes the sum of elements on the main diagonal. Note that the residual orientation angle is symmetric, $\epsilon_{\mathbf{\Delta R}}(c,d) = \epsilon_{\mathbf{\Delta R}}(d,c)$, while the residual displacement length is in general not, $\epsilon_{\mathbf{\Delta t}}(c,d) \neq \epsilon_{\mathbf{\Delta t}}(d,c)$. This is because the computation of $\mathbf{\Delta}\widehat{\mathbf{t}}^{c\to d}$ depends on $\mathbf{\Delta}\widehat{\mathbf{R}}^{d}$, while the computation of $\mathbf{\Delta}\widehat{\mathbf{t}}^{d\to c}$ depends on $\mathbf{\Delta}\widehat{\mathbf{R}}^{c}$.

The mean residual orientation angle and displacement length across all cameras are then given by

$$\epsilon_{\mathbf{\Delta R}} = \frac{1}{C^2 - C} \sum_{c,d=1...C} \epsilon_{\mathbf{\Delta R}}(c,d), \text{ and} \tag{8.4}$$

$$\epsilon_{\mathbf{\Delta t}} = \frac{1}{C^2 - C} \sum_{c,d=1...C} \epsilon_{\mathbf{\Delta t}}(c,d), \tag{8.5}$$

respectively, where C is the number of cameras[1]. While $\epsilon_{\mathbf{\Delta R}}$ and $\epsilon_{\mathbf{\Delta t}}$ are independent of the chosen reference coordinate frame and parameterization, only $\epsilon_{\mathbf{\Delta R}}$ is also independent of the choice of scale fixing. In the remainder of this thesis we refer to the mean residual orientation angle and displacement length as orientation and displacement error, respectively.

Ground truth intrinsic and extrinsic calibration parameters were acquired in an extensive offline calibration procedure. First, all cameras were calibrated intrinsically. To this end, the calibration toolbox of Mei and Rives [64] was used to acquire an initial set of intrinsic calibration parameters. We augmented this method by estimating the displacement of the projection centers and finally approximated the noncentral camera by a central camera as proposed by Schönbein et al. [86] (cf. Section 3.2.2). Then the cameras were mounted on the test vehicle and the setup was calibrated extrinsically. Coded calibration targets were placed around the vehicle and on the floor, covering large regions of the fields of view. The poses of the calibration targets and cameras were then computed using the camera images, a professional photogrammetry software, and additional images from a hand-held camera.

To detect putative alterations of the setup during data recording the offline calibration procedure was performed twice, before and after recording the dataset. The mean residual orientation angle and displacement length between both calibrations are $0.073°$ and 2.5mm, respectively. We combined both calibrations into a single ground truth calibration using pose interpolation.

8.3 Initialization and Parameter Tuning

To generate quantitative results we ran our extrinsic self-calibration algorithm offline using perturbed samples of the ground truth calibration parameters for

[1] Note that $\epsilon_{\mathbf{\Delta R}}(c,c) = 0$ and $\epsilon_{\mathbf{\Delta t}}(c,c) = 0$ for $c = 1...C$.

initialization. A total of 20 samples was drawn prior to the evaluation. To generate the samples each camera was offset 0.5m in a random direction and then rotated through a random angle between 0° and 15° about a random rotation vector. Figure 8.4 illustrates a subset of the drawn samples. The median initial orientation and displacement errors across the 20 samples are 10.3° and 700.9mm, respectively. Since the parameters of each camera were perturbed individually, the relative orientation error between adjacent cameras may exceed 15°. Furthermore, due to the influence of the orientation error on the displacement error, the median initial displacement error exceeds 0.5m. The dynamic parameters were initialized assuming the vehicle to drive in a straight direction parallel to the ground plane. However, due to perturbation of the orientation of the reference camera parameters, the translation direction which is defined in the coordinate frame of the reference camera is not necessarily parallel to the ground plane. The initial velocity was set to 20 kilometers per hour. The initial a priori covariance matrix and the process noise were manually tuned on the subset of 4 out of the 24 sequences (shown in orange in Figure 8.3). We were aiming at accurate results while ensuring convergence.

8.4 Quantitative Evaluation

In the following we present the quantitative evaluation of our continuous extrinsic self-calibration algorithm. We present results for the two motion models as well as for a combined calibration incorporating image correspondences between overlapping fields of view.
The evaluation is divided into two parts. First, we compare the results of our algorithm directly with the reference calibration by applying the error metric described in Section 8.2. To this end, we initialized the algorithm as described in the previous section. This experiment was repeated 400 times for each configuration. This first experiment provides an application-independent and thus general assessment of the calibration results. In a second experiment, we compare the calibration results against the reference calibration in a typical application, visual odometry. Throughout this section we use box plots to illustrate results. A detailed explanation can be found in Appendix A.8.

8.4.1 Motion-Based Calibration

Motion-based calibrations builds on the rigid coupling between the cameras and in particular on the different apparent motions observed by each camera when the setup is moved. Figure 8.5 shows the evolution of orientation and

(a) Side view of ground truth camera poses and subset of initialization samples.

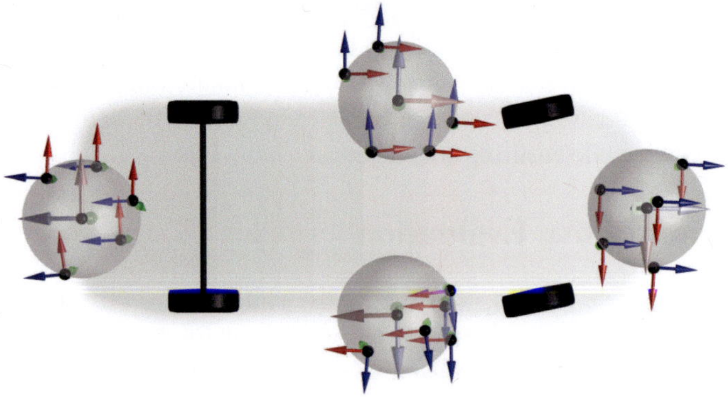

(b) Top view of ground truth camera poses and subset of initialization samples.

Figure 8.4: Ground truth camera poses and a subset of drawn samples for initialization are shown in a side view (a) and in top view (b). Large coordinate axes indicate the ground truth camera poses. Smaller coordinate axes visualize a subset of the initialization samples which are offset by 0.5m and rotated through an angle of up to $15°$ with respect to the ground. For reference, transparent spheres with 0.5m radius along with vehicle tires and the rear axle are shown.

displacement residuals and various other parameters over time for one exemplary calibration run. During calibration the errors are reduced from initially $11.82°$ and 839.5mm to $0.24°$ and 30.8mm, respectively. While the orientation error decreases monotonically to a low value, the displacement error first settles at approximately 670mm and then decreases to approximately 100mm within 50 frames. The reason for this behavior is the first of four turns. In Chapter 4 we have shown that observability of the parameters depends on the type of motion. Since the vehicle was driving straight in the beginning some parameters remained unobservable during this time.

Furthermore, the displacement error fluctuates strongly within the first 100 frames. This typical behavior is caused by the concurrent update of several parameters. A behavior similar to that of the displacement residual can be observed for the estimated height.

For the quantitative evaluation, the calibration algorithm was tested on all combinations of the 20 evaluations sequences and 20 initial parameter samples, yielding 400 runs per configuration. We define the calibration result as the current parameter estimate at the end of each sequence. Figure 8.6 shows the results for the planar and general motion model. The results for the general motion model show higher variance but a lower orientation error and a similar median displacement error. The median values of the initial errors at start-up are $10.3°$ and 700.9mm. After calibration, the median values are $0.22°$ and 36.5mm for the planar motion model, and $0.17°$ and 35.6mm for the general motion model. Hence, the self-calibration algorithm was able to reduce the median orientation error by a factor of 50 and the median displacement error by a factor of roughly 20. However, the algorithm does not always converge to the correct solution. The number of data points not shown in the box plots are given in Appendix A.9. We conclude that the performance of both models is similar, with general motion model having a lower median orientation error but higher displacement error variance.

Due to the different error metrics, we cannot compare our results directly with those of Pagel et al. [73, 75] and Heng et al. [36, 35].

For evaluation, Pagel et al. [73, 75] use a setup consisting of three cameras which are assumed to be coplanar, thus estimating only a subset of the extrinsic calibration parameters. Unfortunately, numeric extrinsic calibration results are not provided. Pagel et al. report an average orientation error of $0.8°$ and an average displacement error of 10.8%. The average baseline of our setup is around 2.8m. Hence, a 10.8% error corresponds to approximately 30cm.

Heng et al. [36, 35] provide numeric results on the residual orientation and displacement angle between the front, reference camera ($r = 0$) and the other

Approach	Orientation error	Displ. angle error
[36]	$0.87°$	$1.99°$
[35]	$0.43°$	$1.47°$
(two vehicles)	$0.41°$	$1.57°$
Planar model	$0.15°$	**0.43°**
General model	**0.11°**	$0.46°$

Table 8.1: Comparison of the results of our motion-based calibration approach (planar and general) with the results of Heng et al. [36, 35]. Note that the results of our approach are median values of 400 runs, respectively, while the results of Heng et al. are single run results.

cameras $c \in \{1, 2, 3\}$. We can compare our results against those of Heng et al. by computing the mean orientation error and mean displacement angle error (angle between $\mathbf{\Delta t}^c$ and $\widehat{\mathbf{\Delta t}^c}$) for a given setup. The results are shown in Table 8.1. Note that this error metric is not independent of choice of reference camera and both approaches of Heng et al. are offline calibration methods. However, the authors found the results obtained with respect to a reference calibration to be inconclusive, since the proposed methods yielded better results than the reference method in qualitative validation experiments.

Varying the Frame Rate

The data in our dataset was recorded at 30Hz. By subsampling the image stream we can simulate lower frame rates. Figure 8.7 shows the results of the motion-based calibration algorithm at reduced frame rates. We can see that reducing the frame to 10Hz has only minor impact on the calibration results. In fact, the median orientation error remains within a range of $0.06°$ for all shown results. However, the displacement error increases dramatically at lower frame rates. This applies in particular to the general motion model. Results for the general motion model at 5Hz are not shown here since an appropriate presentation was not possible without rescaling the axes. At lower frame rates (or higher velocities), using a planar motion model is therefore advisable.

8.4.2 Overlapping Fields of View

In the following we present the results of the extrinsic self-calibration algorithm using additionally overlapping fields of view. Figure 8.8 shows the results for the combined calibration. The median orientation error is 0.1° for both motion models, and the median displacement error is 16.2mm and 10.8mm for the planar and general motion model, respectively. These results are substantially better then those obtained from motion based calibration.
In [43] it was shown that the same experimental setup can be calibrated solely based using overlapping fields of view, without the need for motion-based calibration. For this reason, we performed an additional experiment using only the overlapping fields of view between the left and backward-facing camera. The results are shown in Figure 8.9. The median orientation and displacement error in this case are 0.16° and 30.2mm for the planar motion model, and 0.13° and 31.9mm for the general motion model, respectively. In both cases we observe an improvement over motion-only calibration.
Applying the same metric that was used to compare our results with those of Heng et al. [36, 35] (cf. Section 8.4.1), we achieve an orientation error and displacement angle error of 0.09° and 0.09°, respectively, using the combined calibration with the planar motion model, and 0.08° and 0.10°, respectively, using the general motion model.

8.4.3 Visual Odometry Loop Closure Error

The sensitivity of an application output with respect to errors in the individual calibration parameters is in general application dependent. In a virtual top view application, for example, in-plane displacement errors result in a shift in the top view image of the same amount. However, the shift caused by an error in height depends on the angle of incidence and becomes much larger at shallow angles. Here, we use visual odometry to assess the calibration results in the context of an exemplary application.
Visual odometry is the process of estimating camera motion from images only. Due to errors in the calibration, measurement noise, and violations of the motion model and Kalman filter assumptions, odometry errors will accumulate over time. We use the accumulated error, i.e. the drift, as a measure to assess the calibration results. Since the test vehicle is not equipped with sensors that allow to determine its pose directly with high precision, we used manually selected image correspondences to compute the relative pose of the multi-camera system between different time instances. To this end, we selected image pairs which

have been captured from a similar place and with a similar vehicle pose, e.g. at the start and end of each sequence (cf. Figure 8.3). This process is illustrated in Figure 8.10). First, 3D points are triangulated using image correspondences from the same time instance (bottom right). An initial relative pose estimate is then computed by aligning the triangulated 3D point positions. The final estimate is obtained by refining the initial estimate using all correspondences. This estimate is then treated as ground truth during the evaluation. We used the visual odometry algorithm which is described in the next section to estimate the vehicle trajectory. The residual rotation and translation are shown in Figure 8.11. We ran the visual odometry algorithm 40 times on each of the 20 evaluation sequences using one randomly drawn calibration result out of the 400 previously computed results each time. In total, 800 runs were conducted for the motion-based and combined calibration, respectively, and 20 runs were conducted using the ground truth calibration. As some sequence contain multiple loop-closures the number of data points in Figure 8.11 exceeds the number of algorithm runs. To account for the different track lengths, the residuals are normalized with respect to the driven distance. We observe that the combined calibration yields substantially better results compared to the motion-based calibration. Furthermore, the results of the combined calibration are close to the results of the ground truth calibration.

8.4.4 Assessing Calibration Results at Runtime

So far we compared the calibration results against ground truth obtained using offline methods. In a typical application, however, this data is not available and we have to rely on the estimates and observations themselves to assess the calibration results. This is necessary since subsequent applications (presumably) rely on a calibrated system. In the following we discuss several approaches to assessing the calibration results at runtime.

Offline calibration methods commonly analyze the measurement residual (after calibration/optimization) [105, 53, 93]. Large residuals may indicate that the current estimate is far from the optimum but could also be caused by violations of the underlying assumptions such as rigidity of the cameras setup. In a calibrated system the remaining measurement residual should be influenced predominantly by the inaccuracy of the feature detection and matching algorithm. Given a calibrated system, we can thus determine typical values which can later be used for comparison. However, low measurement residuals do not necessarily indicate that the system is well calibrated. For example, if the vehicle is only driving straight or not moving at all motion-based calibration is not possible, yet the measurement residuals may be small. This problem is directly related to observability analysis which we discuss next. We conclude that a low measurement residual is a necessary condition for a calibrated system.

To determine whether the calibration parameters can estimated unambiguously from the observations offline calibration methods typically analyze the covariance matrix of the estimated parameters. Given the measurement covariance matrix and assuming the estimation problem not to be over-determined, an approximation of the covariance matrix can be computed through backward propagation [31]. Before elaborating observability analysis we present one way to visualize the uncertainty associated with the current estimate.

In Section 4.1 we discussed system parameterizations and argued that a minimal parameterization is advantageous during optimization, yet other parameterizations such as the free net adjustment might be favorable for analysis. Herein, we use free net adjustment to visualize the uncertainty in the relative displacements between cameras. The camera positions are parameterized by 3-vectors. To compensate for the over-parameterization seven linear constraints are introduced that fix the datum. The constraints correspond to the first order approximation of a similarity transformation that minimizes the mean Euclidean distance between the current camera position estimates and initial camera position estimates[2]. To obtain the covariance matrix of the camera positions we apply forward propagation of the a posteriori covariance matrix. This is illustrated in Figure 8.12. We observe that the relative camera heights can be estimated with higher accuracy than the in-plane displacements. A drawback of this representation is that the correlation between orientations cannot be visualized.

If the vehicle was driving only straight ahead, the in-plane camera displacement cannot be observed by means of motion-based calibration (cf. Section 4.2). In this case we expect the corresponding entries in the covariance matrix to be very large[3]. Depending on the parameterization it might not be easy to determine if covariance values are uncommonly high (due to different units and ranges) and understand the physical implications (due to correlations). For example, while we describe the relative displacement between the forward and side-facing cameras using 3-vectors we use two angles and a fixed distance for the backward-facing camera. A common way to account for the different units and ranges is to normalize (whiten) the covariance matrix with respect to initial estimate or

[2] In the following example we used the ground truth position instead of initial the estimates for visualization.

[3] Due to measurement noise, nonlinearities, and other error sources the values will not be infinite.

system noise covariance matrix [29]. The resulting covariance matrix is dimensionless and normalized. Then we compute the eigenvectors and eigenvalues of the matrix. The eigenvector corresponding to the highest eigenvalue indicates the direction in parameter space with the highest uncertainty. The corresponding eigenvalue corresponds to the variance in this direction. For the case of the straight driving vehicle in the above example we expect two similarly large eigenvalues.

This method is promising in case where the a priori and process noise covariance matrices are physically motivated, e.g. by long term drift analysis of similar systems, and the system is linear. However, here we use pseudo-noise ([6]) for both covariance matrices to control the behavior of the extended Kalman filter.

Figure 8.13 shows the evolution of the displacement error and the estimated uncertainty over time. As expected the variance decreases after the first turn and increases while driving straight (magnified view). However, we observe that the extended Kalman filter severely underestimates the covariance (the decrease in the displacement error and standard deviation differ by a factor of more than five). This property of the extended Kalman filter is a well-known ([61]) and caused by nonlinearities and model violations.

Finally, heuristic indicators can be used to evaluate whether the algorithm is working as intended. For example, the number of ground plane inliers reflects the ability to track the ground plane and thus to estimate the relative orientation parameters. Furthermore, we can analyze the estimated trajectory with respect to straight driving and turning maneuvers. A well-established approach is to conduct a control experiment. For example, we could detect single distinct features and test whether they can be observed again (at the same or at a later time) at the expected position in the field of view of another camera. We can also take advantage of the continuous stream of observations and analyze the a posteriori measurement residual during short periods in which the calibration parameters are fixed, estimating only the dynamic parameters. Maye et al. [63] propose accumulating a small representative set of observations, e.g. several consecutive pairs of frames, which can be used to test the current estimate.

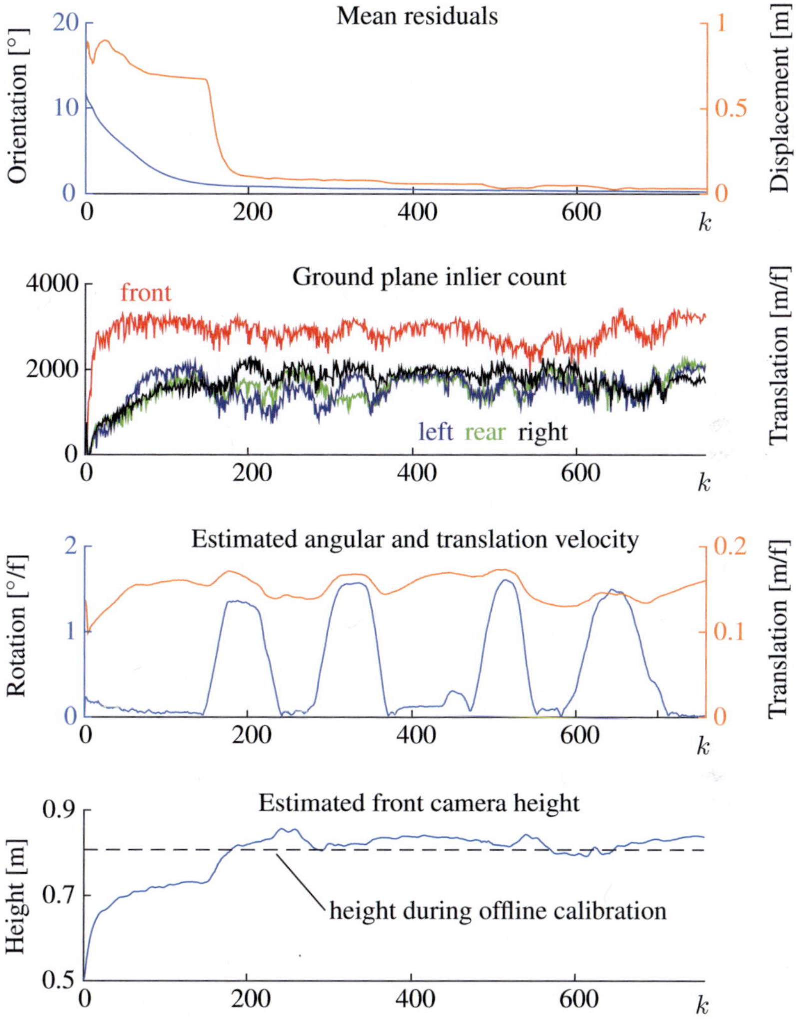

Figure 8.5: Evolution of residuals, ground plane inliers, and estimated quantities over time for one exemplary calibration run. The top plot shows the mean orientation (blue) and displacement residuals. Below, the evolution of the detected ground plane inlier correspondences is shown for the front (red), left (green), rear (blue), and right-facing camera (black), respectively. Next, the estimated rotation angle (blue) and translation length (orange) per frame are shown. The bottom plot shows the estimated height of the front camera. The dashed line depicts the camera height during offline calibration.

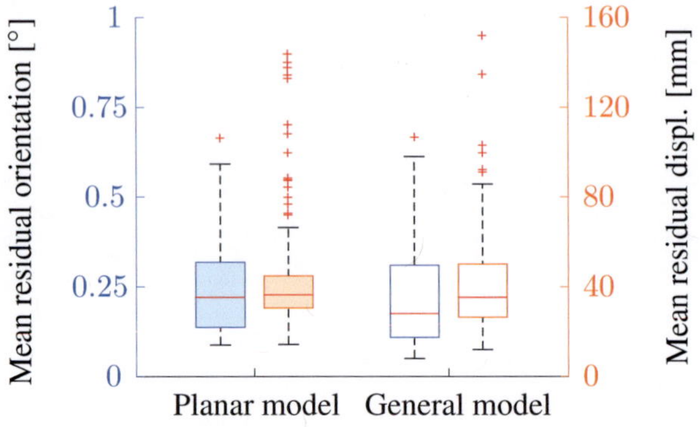

Figure 8.6: Results for motion-based extrinsic self-calibration. The mean residual orientation angles (blue) and displacement lengths (orange) are shown for the planar motion model (filled boxes) on the left and for the general model on the right. Each column represents the results of 400 algorithm runs (20 sequences, 20 initializations). The median values are 0.22° and 36.5mm and 0.17° and 35.6mm, respectively. For a discussion refer to Section 8.4.1.

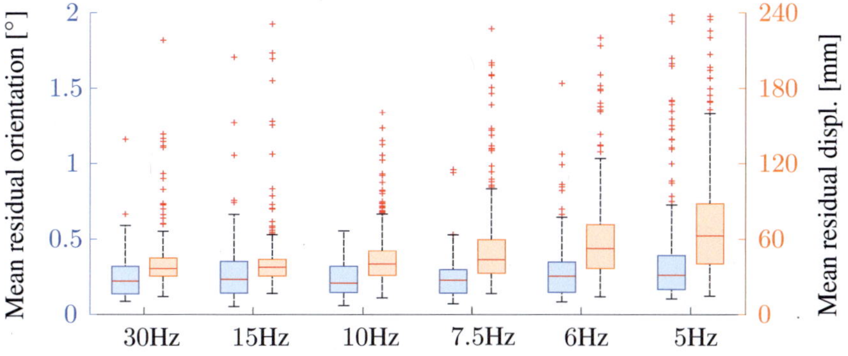

(a) Results for the planar motion model at different frame rates.

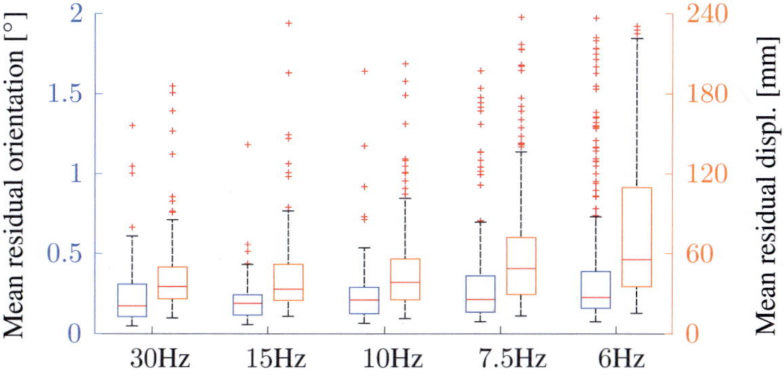

(b) Results for the general motion model at different frame rates.

Figure 8.7: Results of motion-based extrinsic self-calibration at reduced frame rates for the planar motion model (top) and general motion model (bottom). Box plots for equal frame rates are aligned. For reference, the results shown in Figure 8.6 are shown here again. However, note that the axes are scaled differently. Results for the general motion model at 5Hz are not shown since an appropriate presentation was not possible without rescaling the axes. The median values for all box plots are given in Appendix A.9.

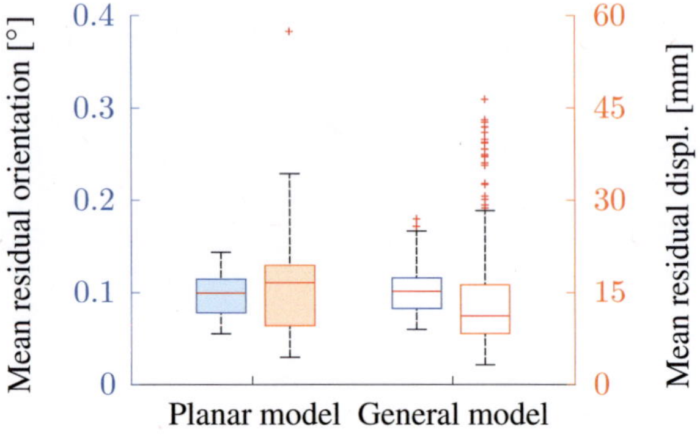

Figure 8.8: Results for the combined calibration, using additionally overlapping fields of view. The median values are $0.1°$ and 16.2mm and $0.1°$ and 10.8mm, respectively.

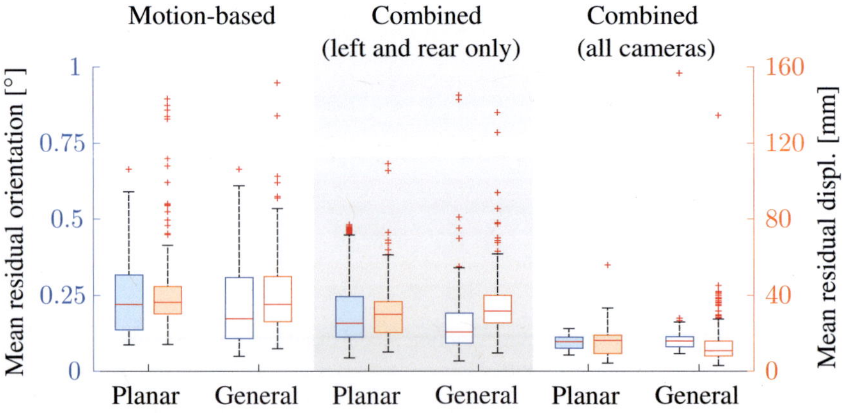

Figure 8.9: Comparison of calibration results between motion-based calibration (cf. Figure 8.6), calibration incorporating all overlapping fields of view (cf. Figure 8.8, and incorporating only the overlapping fields of view of a single camera pair.

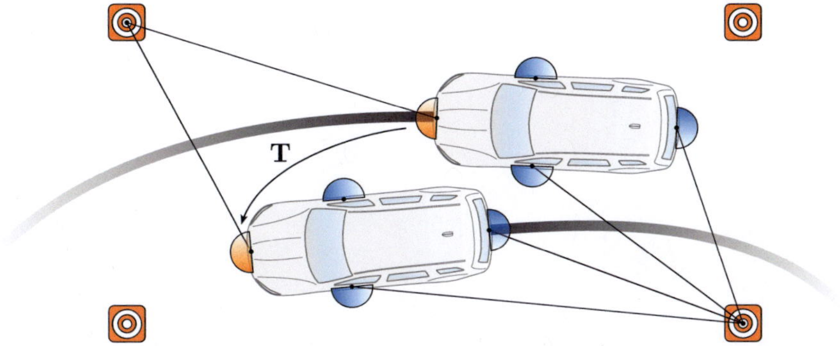

Figure 8.10: Estimating the relative vehicle pose $\mathbf{T}$ between start and end. The relative pose is estimated using manually selected image correspondences. A subset is shown as line segments. To mark the start and end and to simplify the process of determining image correspondences, traffic cones have been placed around the vehicle at the start.

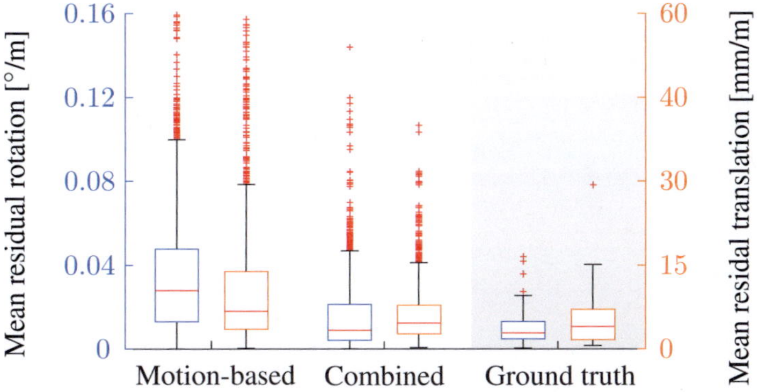

Figure 8.11: Residual rotation and translation per meter traveled for our visual odometry algorithm using calibration results from motion-based calibration, combined calibration using overlapping fields of view and ground truth calibration. Note that the number of data points varies significantly (see text). Median values and number of not shown data points are given in Appendix A.9.

Figure 8.12: Exemplary camera center covariance ellipsoids (gray, rescaled) for one result of motion-based calibration, in a top view (left) and perspective view (right). Free net adjustment is used to propagate the state covariance matrix (see text). The mean Euclidean distance between the cameras position of the current estimate and ground truth are minimized.

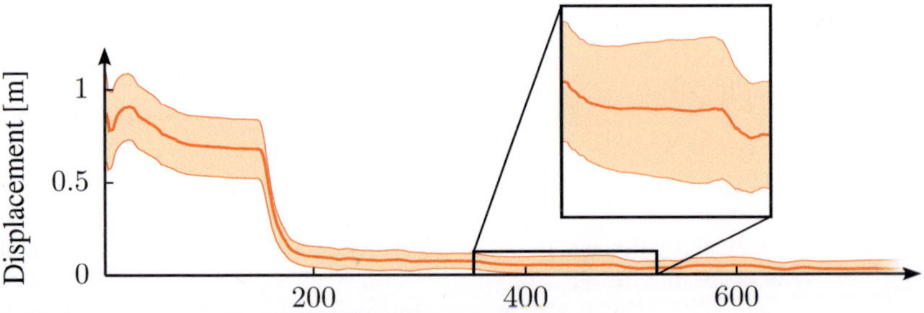

Figure 8.13: Evolution of the displacement error and corresponding (scaled) estimated variance over time. The state covariance was propagated assuming the current estimate to coincide with the ground truth. For reference, the data is offset using the real displacement error (cf. Figure 8.5). The magnified view shows a section of straight driving during which the variance increases.

8.5 Qualitative Results

In the remainder of this chapter we show three typical applications of vehicle-mounted multi-camera systems. Estimated extrinsic calibration parameters were used in all three cases.

8.5.1 Visual Odometry

Visual odometry is the process of estimating the motion of a camera system from images only. The estimated trajectory can be used for vehicle navigation or 3D reconstruction tasks. Several different approaches for visual odometry using monocular and stereo cameras (e.g. [71]), as well as multiple cameras without overlapping fields of view (e.g. [41]) have been proposed. Herein, we present results which are based on the motion estimates of the extended Kalman Filter used for extrinsic calibration. At each time step, the filter provides an incremental motion estimate, $\widehat{\mathbf{T}}_k^r$, that relates the current and preceding pose of the reference camera. From this, the trajectory of the reference camera is obtained through concatenation. To visualize the results we reconstruct the ground plane texture using estimates of the camera motion and ground plane. Locations that are passed multiple times by the vehicle such as start and end are reconstructed multiple times. Hence, errors in the motion estimates will cause ghosting artifacts, i.e. the same texture will appears multiple times with offsets. These artifacts can be used as a simple way to assess the estimation results.
For the reconstruction, we initialize the extended Kalman filter with an earlier calibration result. During the reconstruction the extrinsic calibration is then kept constant by adjusting the initial state covariance and process noise. The ground plane texture is reconstructed incrementally using only the motion and ground plane estimates $\widehat{\mathbf{T}}_k^r$, $\widehat{\mathbf{n}}_k^r$, and $\widehat{h}_k^r$, that are available up to the current point in time, thus enabling online processing.
In general, the ground plane and motion estimates are not consistent over time, i.e.

$$\widehat{\mathbf{n}}_{k+1}^r \neq \widehat{\mathbf{R}}_k^r \widehat{\mathbf{n}}_k^r. \tag{8.6}$$

To achieve consistency we favor the current ground plane estimates over results obtained through concatenation. Figure 8.14 shows the reconstructed ground plane texture for one of our test sequences. For motion and ground plane estimation every second frame was skipped to minimize drift. All visual odometry approaches suffer from the fact that introduced errors cannot be corrected later on, inevitably causing the trajectory to drift (cf. Figure 8.14).

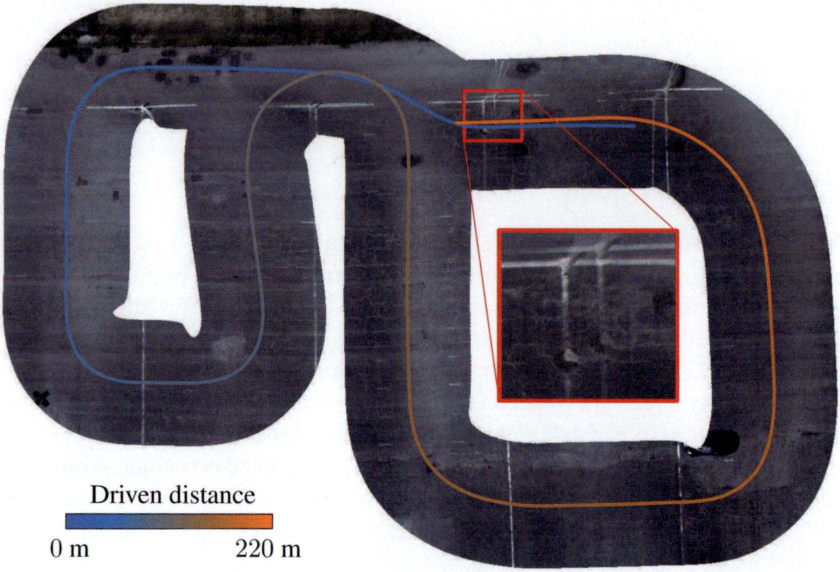

Figure 8.14: Ground plane texture reconstruction for one of the test sequences captured in a parking area. The image was generated incrementally from 720 images of the the backward-facing camera by reconstructing and blending semicircular ground plane texture patches. The estimated vehicle trajectory is shown as the colored curve. The color indicates the estimated traveled distance. At locations which were passed twice ghosting artifacts can be observed. A magnified view of a ghosting artifact at sequence end is shown. The offset is roughly 0.85 meters corresponding to 0.4 percent of the traveled distance. Note the reconstruction of the parked vehicle in the bottom right corner.

Additionally, special motions such as linear or circular motions cause further drift since camera velocity and height cannot be observed.

8.5.2 Virtual Top View

A classic application of vehicle-mounted multi-camera systems is the generation of a virtual top view. To this end, the images of the four cameras are projected onto the ground plane and blended into a composite image. Either static or dynamic ground plane parameters can be used for this purpose. Similar to the visual odometry approach in the previous section we used the extended Kalman filter to estimate the ground plane parameters dynamically. The extrinsic calibration parameters were initialized using an earlier calibration result and then kept constant by adjusting the initial state covariance and process noise. Figure 8.15 shows two examples of generated virtual top view images. Dynamically adjusting the ground plane parameters allows to compensate for nonplanar vehicle motions such as rolling and pitching. Even during a tight turn the system does not create visible ghosting artifacts (see Figure 8.15b).
The low camera height and large extent of the virtual top view image (cf. Figure 8.16) cause inhomogeneity in spatial resolution. Furthermore, the assumption of the vehicle surrounding being planar causes all nonplanar objects to appear significantly distorted, e.g. the vehicles in Figure 8.15a. This effect could be mitigated by mounting the cameras in a higher position or by applying a more sophisticated surface model.

8.5.3 Stereo Rectification

If the intrinsic and extrinsic calibration parameters of a pair of cameras are known and the cameras share a common field of view it is possible to reconstruct 3D points in the scene from image correspondences. The two-dimensional search for image correspondences can be reduced to one dimension by applying stereo rectification. In Section 3.3.1 it was shown that the ray corresponding to an image point and the displacement vector between the cameras define the epipolar plane (see Figure 8.17a). The intersections of the epipolar plane with the image planes define the epipolar lines. The corresponding image points have to be located on the epipolar lines. In general, the epipolar plane will be imaged as a curve due to nonlinearities in the imaging process. The goal of stereo rectification is to simplify the correspondence search by warping the images of both cameras such that epipolar curves are mapped to parallel lines. It is common to apply a mapping that cause the epipolar lines to be parallel to the u-axis and match across images.
In this case, given a point $\mathbf{u} = (u, v)^T$ in the rectified image of the first camera the corresponding point in the rectified image of the second camera has coor-

(a) Virtual top view image of the vehicle driving straight.

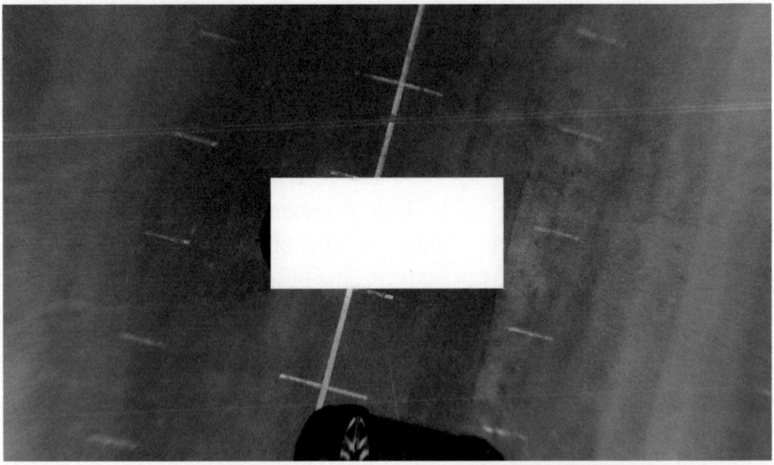

(b) Virtual top view image of the vehicle turning.

Figure 8.15: Two virtual top view images. The images were generated by projecting the image of the four cameras onto the ground plane and apply image blending. The masks used for blending and the dimensions of the virtual top view images are shown in Figure 8.16. By adjusting the ground plane normal dynamically ghosting artifacts introduced by rolling and pitching of the vehicle can be compensated. The vehicle was driving to the left in the first image and making a tight turn in the second image.

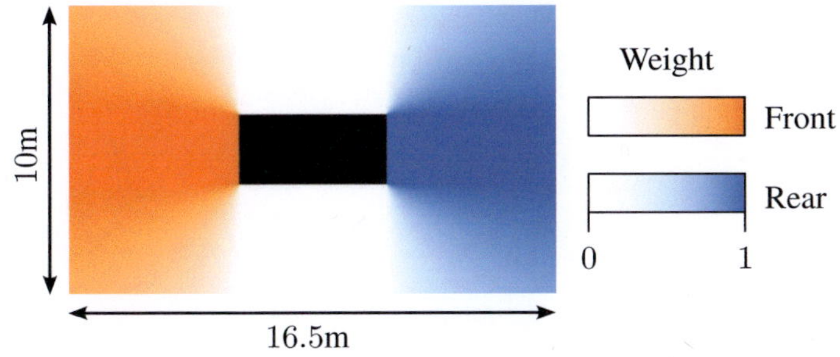

Figure 8.16: Blending mask for the front and backward-facing camera, respectively, and dimensions of the reconstructed ground plane region. The images from the four cameras are blended into a composite image by applying a blending mask with fixed weights. The weights of the forward and backward-facing camera are shown here exemplary. The texture beneath the vehicle cannot be reconstructed and is shown in black.

dinates $\mathbf{u}' = (u', v)^T$. Hence, the stereo rectification simplifies the correspondence search along a parametric or nonparametric curve to a search along a horizontal line. In general, the search space can be further reduced by taking into account that the 3D point has to be located between the image of the projection center of the first camera and infinity.

Stereo rectification consists implicitly of two steps. In the first step the cameras are virtually rotated around their respective camera centers such that the principal axes are parallel to each other and perpendicular to the displacement vector. Typically, another rotation is applied to align X_C-axes of the camera coordinate frames[4]. This process is illustrated in Figure 8.17b. In the second step, a new camera projection model is applied. The new model has to satisfy the above constraint of projecting the epipolar planes onto parallel (and matched) lines. The pinhole model satisfies this constraint. If the same model parameters are chosen for both cameras and the epipolar lines are matched across the images, the disparity $u - u'$ is proportional to the inverse of the depth of the 3D point. Here we define depth as the distance along the principal axis of the virtually rotated camera.

[4] Note that there is one degree of freedom corresponding to the rotation about the displacement vector.

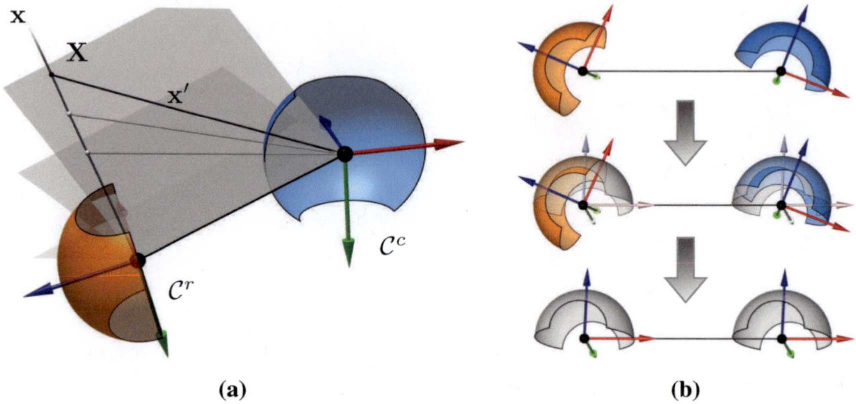

Figure 8.17: Epipolar geometry (a) and camera alignment for stereo rectification (b). The displacement vector between the two cameras $\mathcal{C}^r$ and $\mathcal{C}^c$ and an arbitrary 3D point $\mathbf{X}$ define an epipolar plane (several planes shown here). For stereo rectification, the cameras are virtually rotated such that the principal axes are parallel and the x-axes are collinear with the displacement vector (b).

For fisheye cameras different projection models for stereo rectifications are preferred due to the large field of view (cf. Section 3.2.1). For example, Abraham and Förstner [1] propose two models, a stereographic rectification model and an equidistant rectification model. Here we present results for the latter.
Figure 8.18 shows one image from the front-facing camera and a simultaneously captured image from the right-facing camera. Matching epipolar curves are superimposed. In contrast to a perspective camera, here the image disparity is proportional to the intersection angle between the rays in case of the equidistant rectification model. One easily verifies that the topmost epipolar curves both intersect the top of the lamp post. Figure 8.19 shows a corresponding rectified image pair.

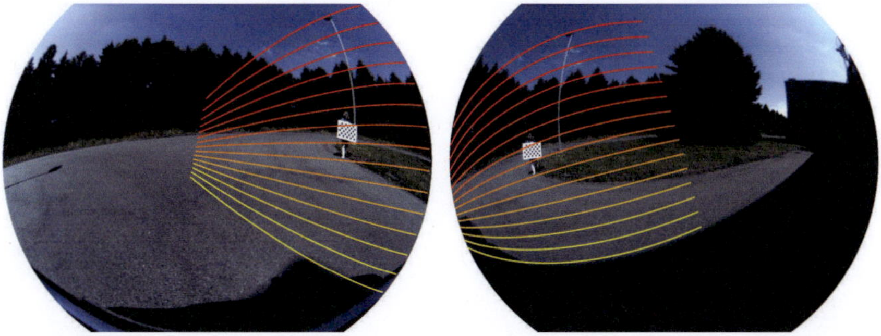

Figure 8.18: Simultaneously captured images from the front (left) and right-facing (right) cameras, respectively. Epipolar curves are superimposed. Matching curves have the same color. The curves are truncated to valid ranges. The relative pose between the cameras was estimated using our extrinsic self-algorithm algorithm. The configuration is the same as the one shown in Figure 8.17.

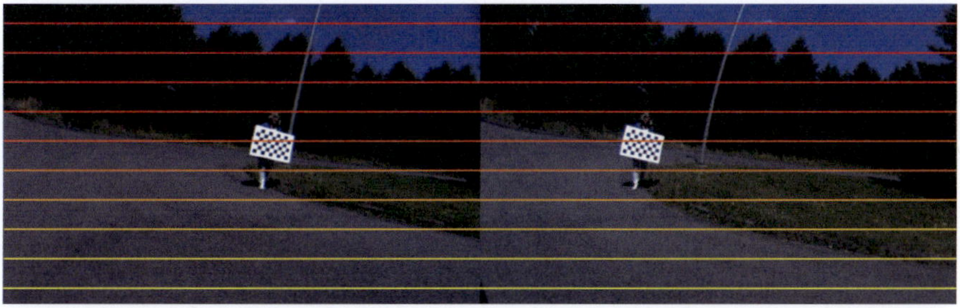

Figure 8.19: Stereo rectified image pair corresponding to Figure 8.18. For rectification the equidistant model [1] was used. One can observe that, e.g. the same tiles of the calibration target are intersected by the epipolar line. The wide baseline causes the calibration target to obstruct parts of the lamp post in the left image while being considerably offset in the right image.

9 Conclusion and Future Research Directions

In this thesis we built the theoretic foundation for continuous extrinsic multi-camera self-calibration. In addition, we proposed and evaluated a Kalman filter-based approach which relies solely on image data. The fields of application are mobile robots and road vehicles equipped with multi-camera systems.

Generally, the extrinsic calibration accuracy of any multi-sensor system deteriorates over time due to external influences such as mechanical stress, vibrations, individual sensors being mounted on moving parts, or because it has been inaccurate from the start. Typically, subsequent functions can cope with an inaccurate calibration to some extent but eventually recalibration becomes inevitable. The calibration of multi-camera systems commonly requires expert knowledge and artificial calibration objects and is thus both time consuming and costly. Furthermore, once a system is deployed it might not be accessible anymore. Continuous self-calibration is the process of estimating the calibration parameters from observations made during regular operation. It is the only way to guarantee reliable long-term operation.

We approached the problem of extrinsic self-calibration by analyzing different combinations of vehicle motion types, sensor configurations, motion estimation algorithms, and scene properties with respect to the constraints they impose on the calibration problem. Fundamental to all calibration constraints is the rigidity assumption of the multi-camera setup over time. In fact, rigidity along with overlapping fields of view is sufficient to enable metric calibration in case of two-camera systems. For more than two cameras, however, these conditions are insufficient. We introduced a matrix rank criterion along with two additional necessary conditions that provide a binary observability measure for multi-camera setups with pairwise overlapping fields of view. Furthermore, for motion-based calibration we presented a set of algorithms to recover the subset of non-ambiguous extrinsic calibration parameters, assuming error free measurements. We concluded that general motion provides a sufficient set of constraints for extrinsic calibration. In case of planar motion additional constraints such as those provided by a jointly observed scene plane or overlapping fields of view are required. With regards to future extensions, it remains a challenging research topic to formulate a general framework that given a sensor configuration, type of class of vehicle motion, and scene, provides an observability measure which enables the comparison and improvement of sensor configurations and compositions.

Relative pose estimation is the essence of extrinsic calibration. To facilitate image-based relative pose estimation in wide-baseline scenarios we proposed an image preprocessing step that compensates geometrical distortions introduced by lens distortions and viewpoint changes. To this end, we utilize prior knowledge of the relative cameras poses and make simplifying assumptions about the scene. In particular, we assume the scene to be composed of a ground plane and distant objects only. Following this approach, we were able to successfully match point features in scenarios where more sophisticated and complex methods previously failed. In addition, we were able to skip several frames during processing with only minor influence on the calibration results.

To track the ground plane over time, we introduced a novel ground plane estimation algorithm for fisheye cameras which is designed to be robust with respect to sparse outliers among putative image correspondences as well as structural outliers such as other planes in the scene. It relies on a sequential updating scheme that favors correspondences that exhibit a high probability of being classified correctly. Correspondences which are found to be induced by a ground plane homography are used to update the estimate, thus facilitating subsequent classification.

The algorithm was integrated into an extended Kalman filter for continuous extrinsic self-calibration. The state vector of the Kalman filters comprises only the calibration parameters, the vehicle dynamics, and ground plane and has thus a low dimensionality compared to approaches that perform structure computation, e.g. [75]. The parameter update is carried out sequentially. First, putative image correspondences are computed using standard feature detection and matching algorithms as well as image prewarping. Inliers among putative correspondences are identified using random sampling consensus. The inliers are then used to update the state vector and covariance matrix. The result is then further processed by the ground plane tracking algorithm which identified ground plane induced correspondences on the basis of the partially updated state vector and an updating scheme that favors high-confidence inliers.

We evaluated the proposed extrinsic self-calibration algorithm using a vehicle-mounted multi-camera setup consisting of four fisheye cameras. In a quantitative evaluation we compared results based on a planar and general motion model and optionally overlapping fields of view directly against a reference calibration. In

an additional experiment, we compared the calibration results against the reference calibration using visual odometry, which represents a typical application. Initial parameters were generated by adding a random displacements of 0.5m and rotating each camera by up to 15° about random rotation axes. Applying our motion-based extrinsic calibration algorithm, we were able to reduce the median initial displacement and orientation errors by a factor of 20 and 50, respectively, from 700.9mm and 10.3° to 36.5mm and 0.22° using the planar motion model, and 35.6mm 0.17° using the general motion model. When incorporating overlapping fields of view, we were able to further reduce the errors by a factor of 2 to 3, down to 16.2mm and 0.1° using the planar motion model, and 10.8mm and 0.1° using the general motion model.
Visual odometry performed similarly on the reference calibration and calibration based on overlapping fields of view. Throughout the evaluation we observed that the general motion model provides slightly better results than the planar motion model. However, the planar motion model seemed to be more robust as it provided much better results at low frame rates. The remaining median errors are in the order of 0.1° to 0.25° and 10mm to 40mm, with the latter corresponding to less than one percent of the largest baseline in the test setup. Finally, we presented some qualitative results using three typical applications for multi-camera systems, namely visual odometry, a virtual top view, and stereo rectification. During our experiments the algorithm diverged in around 1% of cases. We discussed various approaches to detect such cases as well as degenerate motions but did not conduct any further experiments on this topic.
We also want to mention the runtime of the algorithm. Currently, the algorithm does not run in real-time which is mainly due to the implementation of the prewarping algorithm. A significant speed-up could be obtained by utilizing more suitable hardware for this task such as a graphics processing unit or by computing corresponding image points on dedicated hardware. In addition, using an information filter instead of a Kalman filter would allow for distributed computing. In this work we avoided computation of the scene structure for the most part, mainly for complexity and robustness reasons. However, it is to be expected that approaches estimating camera motion along with scene structure are likely to outperform our approach in most scenarios. Such approaches are typically realized using a decentralized solution in which the motion of each camera is estimated independently to ensure the consistency between estimated camera motion and structure [73, 75]. It is thus required that motion and ground plane estimation can be performed robustly for each camera individually. Centralized approaches (such as ours) avoid this drawback.

An interesting direction of future research is the simultaneous estimation of camera extrinsics and intrinsics. While there exists extensive work on the calibration of standard cameras in the computer vision field, we found few approaches for continuous intrinsic self-calibration of wide-angle cameras. This field is particularly challenging, since it introduces a new class of degenerate cases. However, it is also of great significance since we rely on accurate intrinsic parameters for extrinsic calibration.

Another interesting direction is the integration of other sensor modalities such as an inertial measurement unit which could simplify the estimation substantially. The Kalman filter provides an excellent basis for this type of application.

A Appendix

A.1 Constructing Orthonormal Matrices from Two Vectors

Given two 3-vectors $\mathbf{a}_0$ and $\mathbf{a}_1$, with $\mathbf{a}_1 \neq \mathbf{0}_{3\times1}$, $\mathbf{a}_2 \neq \mathbf{0}_{3\times1}$, and $\mathbf{a}_0 \times \mathbf{a}_1 \neq \mathbf{0}_{3\times1}$, we define a third (orthogonal) vector $\mathbf{a}_2 = \mathbf{a}_0 \times \mathbf{a}_1$. Using the three vectors a rotation matrix $\mathbf{R}_{\mathbf{a}_0,\mathbf{a}_1}$ is constructed by applying Gram-Schmidt orthonormalization on the vectors and concatenating the resulting vectors to a 3×3 matrix. The three unit vectors $\mathbf{v}_0$, $\mathbf{v}_1$ and $\mathbf{v}_2$ are computed as

$$
\begin{aligned}
\mathbf{v}_0 &= \frac{\mathbf{a}_0}{\|\mathbf{a}_0\|_2} \\
\mathbf{v}_1 &= \frac{\mathbf{a}_1 - \mathbf{v}_0\mathbf{v}_0^T\mathbf{a}_1}{\|\mathbf{a}_1 - \mathbf{v}_0\mathbf{v}_0^T\mathbf{a}_1\|_2} \\
\mathbf{v}_2 &= \frac{\mathbf{a}_2}{\|\mathbf{a}_2\|_2},
\end{aligned}
\tag{A.1}
$$

where $[\cdot]_\times$ was defined in equation (3.11). The computation of the second vector $\mathbf{v}_1$ can be geometrically interpreted as a projection of $\mathbf{a}_1$ onto the plane defined by the normal vector $\mathbf{v}_0$. Since $\mathbf{a}_2$ is already orthogonal to $\mathbf{v}_0$ and $\mathbf{v}_1$, only a normalization has to be applied. The rotation matrix is then given by $\mathbf{R}_{\mathbf{a}_0,\mathbf{a}_1} = [\mathbf{v}_0, \mathbf{v}_1, \mathbf{v}_2]$. The simplification in equations (A.1) with respect to the classical Gram-Schmid orthonormalization only applies in this specific scenario.

A.2 Rodrigues Formula for Rotation Matrices

Let $\mathbf{a}$ be a 3-vector and θ a rotation angle, we define $\mathbf{R}_{\mathbf{a},\theta}$ using the Rodrigues formula for a rotation matrix [31]

$$
\mathbf{R}_{\mathbf{a},\theta} = \mathbf{I}_{3\times3} - \frac{\sin(\theta)\,[\mathbf{a}]_\times}{\|\mathbf{a}\|_2} + \frac{(1-\cos(\theta))\,[\mathbf{a}]_\times^2}{\|\mathbf{a}\|_2^2}. \tag{A.2}
$$

A.3 Instantaneous Center of Rotation

For planar motion, i.e. $\mathbf{r}_k^T \mathbf{t}_k = 0$, and non-zero angular velocity there exists a point $\mathbf{s}_k$ for which $\mathbf{s}_k = \mathbf{R}_k \mathbf{s}_k + \mathbf{t}_k$ and $\mathbf{r}^T \mathbf{s}_k = 0$, i.e. the point $\mathbf{s}_k$ is fixed under the transformation and is located in the plane defined by the rotation axis direction and the origin. This point is called the instantaneous center of rotation. It can be found by solving the equation system

$$\begin{bmatrix} \mathbf{I}_{3\times 3} - \mathbf{R}_k \\ (\mathbf{r}_k)^T \end{bmatrix} \mathbf{s}_k = \begin{pmatrix} \mathbf{t}_k \\ 0 \end{pmatrix}, \tag{A.3}$$

which is, due to the rank deficiency of $\mathbf{I}_{3\times 3} - \mathbf{R}_k$ of rank three and, thus, yields a unique solution. The instantaneous center of rotation offers an alternative way to represent planar motions.

A.4 Derivation of Equation (6.3)

In the following we derive equation (6.3). The inverse of the homography matrix $\mathbf{H}_k$ is given by

$$\mathbf{H}_k^{-1} = \left(\mathbf{R}_k^T - \frac{\mathbf{t}_k \mathbf{n}_k^T}{h_k}\right)^{-1} = \left(\mathbf{R}_k^T + \frac{\mathbf{R}_k^T \mathbf{t}_k \mathbf{n}_{k+1}^T}{h_{k+1}}\right) \tag{A.4}$$

$$= \left(\mathbf{R}_k^T + \frac{\mathbf{R}_k^T \mathbf{t}_k \mathbf{n}_k^T \mathbf{R}_k^T}{h_k - \mathbf{n}_k^T \mathbf{R}_k^T \mathbf{t}_k}\right), \tag{A.5}$$

where equation (A.4) follows from

$$\mathbf{T}_k^{-1} = \begin{bmatrix} \mathbf{R} & \mathbf{t} \\ \mathbf{0}_{3\times 1}^T & 1 \end{bmatrix}^{-1} = \begin{bmatrix} \mathbf{R}^T & -\mathbf{R}^T \mathbf{t} \\ \mathbf{0}_{3\times 1}^T & 1 \end{bmatrix}, \tag{A.6}$$

and equation (A.5) follows from equation (4.5) and equation (4.7). With this given, we can derive equation 6.3,

$$\widetilde{\mathbf{H}}_k \mathbf{H}_k^{-1} = \left(\mathbf{R}_k - \frac{\mathbf{t}_k {\mathbf{n}_k}^T}{h_k + \Delta h}\right)\left(\mathbf{R}_k - \frac{\mathbf{t}_k \mathbf{n}_k^T}{h_k}\right)^{-1} \tag{A.7}$$

$$= \left(\mathbf{R}_k - \frac{\mathbf{t}_k {\mathbf{n}_k}^T}{h_k + \Delta h}\right)\left(\mathbf{R}_k^T + \frac{\mathbf{R}_k^T \mathbf{t}_k \mathbf{n}_{k+1}^T}{h_k - \mathbf{n}_{k+1}^T \mathbf{t}_k}\right) \tag{A.8}$$

$$= \left(\mathbf{I}_{3\times 3} - \frac{\mathbf{t}_k \mathbf{n}_{k+1}^T}{h_k + \Delta h}\right)\left(\mathbf{I}_{3\times 3} + \frac{\mathbf{t}_k \mathbf{n}_{k+1}^T}{h_k - \mathbf{n}_{k+1}^T \mathbf{t}_k}\right) \tag{A.9}$$

$$= \mathbf{I}_{3\times 3} + \frac{\mathbf{t}_k \mathbf{n}_{k+1}^T}{h_{k+1}}\left(\frac{\Delta h}{h_k + \Delta h}\right), \tag{A.10}$$

where we make use the identity $\mathbf{t}_k \mathbf{n}_{k+1}^T \mathbf{t}_k \mathbf{n}_{k+1}^T = \mathbf{t}_k \mathbf{n}_{k+1}^T \mathbf{n}_{k+1}^T \mathbf{t}_k$. To obtain equation (A.9) we use the identity $\mathbf{R}_k^T \mathbf{R}_k$.

A.5 Extended Kalman Filter

The motion and ground plane parameters, as well as the relative pose parameters are associated with a single state vector of a dynamic system which evolves, corresponding to a discrete time nonlinear stochastic system [6]

$$\boldsymbol{\xi}_k = \mathbf{f}\left(\boldsymbol{\xi}_{k-1}\right) + \mathbf{q}_k. \tag{A.11}$$

The measurements are perturbed by additive zero mean Gaussian noise

$$\mathbf{z}_k = \overline{\mathbf{z}}_k + \mathbf{w}_k, \tag{A.12}$$

where $\overline{\mathbf{z}}_k$ is the error free measurement vector. The terms $\mathbf{q}_k$ and $\mathbf{w}_k$ denote process and measurement noise, respectively. They are assumed to be zero mean, white, mutually uncorrelated, and Gaussian $\mathbf{q}_k \sim \mathcal{N}(\mathbf{0}, \mathbf{Q}_k)$, and $\mathbf{w}_k \sim \mathcal{N}(\mathbf{0}, \mathbf{W}_k)$. The error free observations obey constraints

$$\mathbf{0} = \mathbf{m}\left(\boldsymbol{\xi}_k, \overline{\mathbf{z}}_k\right). \tag{A.13}$$

The state prediction covariance is given by

$$\mathbf{P}_k^- = \mathbf{F}_k \mathbf{P}_{k-1}^+ \mathbf{F}_k^T + \mathbf{Q}_k, \tag{A.14}$$

where

$$\mathbf{F}_k = \left.\frac{\partial \mathbf{f}(\boldsymbol{\xi})}{\partial \boldsymbol{\xi}}\right|_{\boldsymbol{\xi}=\widehat{\boldsymbol{\xi}}_{k-1}^{+}} \tag{A.15}$$

is the derivative of the state transition function at the updated state estimate at time $k-1$. Similarly, the derivative of the measuement prediction with respect to the a priori state estimate is

$$\mathbf{M}_k = \left.\frac{\mathbf{m}(\boldsymbol{\xi},\mathbf{z})}{\partial \boldsymbol{\xi}}\right|_{\boldsymbol{\xi}=\widehat{\boldsymbol{\xi}}_{k}^{-},\ \mathbf{z}=\mathbf{z}_k} . \tag{A.16}$$

The Kalman gain is

$$\mathbf{K}_k = \mathbf{P}_k^{-}\mathbf{M}_k^{T}\left(\mathbf{M}_k\mathbf{P}_k^{-}\mathbf{M}_k^{T} + \mathbf{W}_k\right)^{-1}, \tag{A.17}$$

and the update equations for the state and its covariance are given by

$$\widehat{\boldsymbol{\xi}}_k^{+} = \widehat{\boldsymbol{\xi}}_k^{-} - \mathbf{K}_k\,\mathbf{m}\left(\widehat{\boldsymbol{\xi}}_k^{-},\mathbf{z}_k\right) \tag{A.18}$$

and

$$\mathbf{P}_k^{+} = \left(\mathbf{I} - \mathbf{K}_k\mathbf{M}_k\right)\mathbf{P}_k^{-}, \tag{A.19}$$

respectively.

A.6 Sequential Processing Algorithm

In the following we present the extension of the sequential processing algorithm [6] for extended Kalman filters with implicit measurements constraints. The sequential processing algorithm replaced equations (A.17) to (A.19) of the extended Kalman filter.

If the measurement noise covariance matrix has block diagonal structure we can write it as

$$\mathbf{W}_k = \operatorname{diag}\left(\mathbf{W}_k^{0},\ldots,\mathbf{W}_k^{i},\ldots,\mathbf{W}_k^{N-1}\right), \tag{A.20}$$

where $\mathbf{W}_k^i$ is a square matrix on the main diagonal. Similarly, the measurement vector and derivative of the measurement constraints have the following structure

$$\mathbf{m}\left(\widehat{\boldsymbol{\xi}}_k^-, \mathbf{z}_k\right) = \begin{bmatrix} \mathbf{m}_k^0 \\ \vdots \\ \mathbf{m}_k^{N-1} \end{bmatrix}, \ \mathbf{M}_k = \begin{bmatrix} \mathbf{M}_k^0 \\ \vdots \\ \mathbf{M}_k^{N-1} \end{bmatrix}. \tag{A.21}$$

Initially, the state state and covariance estimate are associated with the a priori estimate and covariance matrix

$$\widehat{\boldsymbol{\xi}}_k^{-1} = \widehat{\boldsymbol{\xi}}_k^-, \ \mathbf{P}_k^{-1} = \mathbf{P}_k^-. \tag{A.22}$$

Then, for each $i = 0 \cdots N-1$ sequential updates are performed. The Kalman gain is given by

$$\mathbf{K}_k^i = \mathbf{P}_k^i \left(\mathbf{M}_k^i\right)^T \left(\mathbf{M}_k^i \mathbf{P}_k^i \left(\mathbf{M}_k^i\right)^T + \mathbf{W}_k^i\right)^{-1}. \tag{A.23}$$

The updated state is given by

$$\widehat{\boldsymbol{\xi}}_k^i = \widehat{\boldsymbol{\xi}}_k^{i-1} - \mathbf{K}_k^i \left(\mathbf{m}_k^i + \mathbf{M}_k^i \left(\widehat{\boldsymbol{\xi}}_k^{i-1} - \widehat{\boldsymbol{\xi}}_k^-\right)\right), \tag{A.24}$$

where the right expression is the linearization of the constraint function evaluated at $\widehat{\boldsymbol{\xi}}_k^{i-1}$ and linearized at $\widehat{\boldsymbol{\xi}}_k^-$. The associated covariance matrix update is given by

$$\mathbf{P}_k^i = \left(\mathbf{I} - \mathbf{K}_k^i \mathbf{M}_k^i\right) \mathbf{P}_k^{i-1}. \tag{A.25}$$

After N updates, the a posteriori state estimate and associated covariance matrix are

$$\widehat{\boldsymbol{\xi}}_k^+ = \widehat{\boldsymbol{\xi}}_k^{N-1}, \ \mathbf{P}_k^+ = \mathbf{P}_k^{N-1}. \tag{A.26}$$

Note that the algorithm only requires the inversion of matrices of the size of $\mathbf{W}_k^i$, which oftentimes are scalars. Furthermore, if the measurement noise covariance matrix does not have block-diagonal structure, the processing steps are identical to the original extended Kalman filter in Section A.6.

A.7 Derivation of Equation (7.8)

Here, we derive equations (7.8) from equation (7.7). The vector $\mathbf{v}$ in equation (7.7) is

$$\mathbf{v} = \mathbf{X}' - \mathbf{R}\mathbf{X} + \mathbf{n}^T\mathbf{X}\frac{\mathbf{t}}{\|\mathbf{t}\|_2}\tau.$$

Next, we substitute

$$\mathbf{X} = \frac{-h\mathbf{x}}{\mathbf{n}^T\mathbf{x}}, \tag{A.27}$$

using equation (3.13) and

$$\mathbf{X}' = \frac{-\mathbf{x}'}{\mathbf{n}^T\mathbf{R}^T\mathbf{x}'}\left(h - \mathbf{n}^T\mathbf{R}^T\mathbf{t}\right) \tag{A.28}$$

$$= \frac{-h\mathbf{x}'}{\mathbf{n}^T\mathbf{R}^T\mathbf{x}'}\left(1 - \mathbf{n}^T\mathbf{R}^T\frac{\mathbf{t}}{\|\mathbf{t}\|_2}\tau\right), \tag{A.29}$$

using additionally equations (4.5) and (4.7). Dividing by $-h$ yields

$$\frac{-\mathbf{v}}{h} = \frac{\mathbf{x}'}{\mathbf{n}^T\mathbf{R}^T\mathbf{x}'}\left(1 - \mathbf{n}^T\mathbf{R}^T\frac{\mathbf{t}}{\|\mathbf{t}\|_2}\tau\right) - \mathbf{R}\frac{\mathbf{x}}{\mathbf{n}^T\mathbf{x}} + \mathbf{n}^T\frac{\mathbf{x}}{\mathbf{n}^T\mathbf{x}}\frac{\mathbf{t}}{\|\mathbf{t}\|_2}\tau. \tag{A.30}$$

After canceling and and reorganisation we obtain equations (7.8)

$$\frac{-\mathbf{v}}{h} = \frac{\mathbf{x}'}{\mathbf{n}^T\mathbf{R}^T\mathbf{x}'} - \mathbf{R}\frac{\mathbf{x}}{\mathbf{n}^T\mathbf{x}} + \left(\mathbf{I}_{3\times 3} - \frac{\mathbf{x}'}{\mathbf{n}^T\mathbf{R}^T\mathbf{x}'}\mathbf{n}^T\mathbf{R}^T\right)\frac{\mathbf{t}}{\|\mathbf{t}\|_2}\tau. \tag{A.31}$$

Note that $\mathbf{x}/\mathbf{n}^T\mathbf{x}$ and $\mathbf{x}'/\mathbf{n}^T\mathbf{R}^T\mathbf{x}'$ correspond to the 3D points $\mathbf{X}$ and $\mathbf{X}'$ normalized by their negative camera height, respectively.

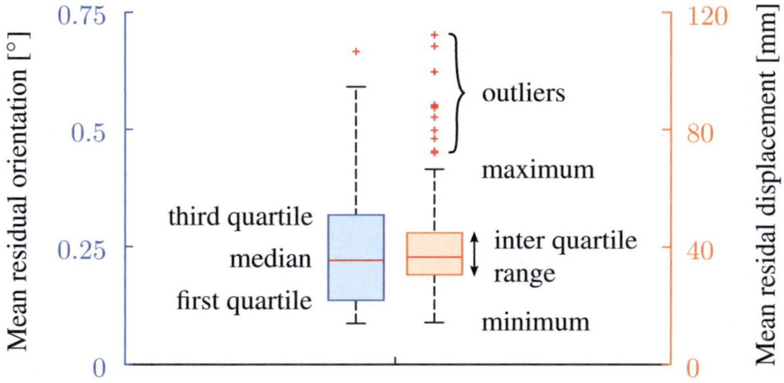

Figure A.1: Example of two parallel box plots. Two vertical axes are used to depict corresponding mean orientation and displacement errors within the same diagram.

A.8 Box Plots

Box plots, or box-and-whisker plots are a simple way to depict data points graphically through their quantiles. An example of two parallel box plots is shown in Figure A.1. In the following, we refer specifically to Tukey box plots [101]. The box plot consists of the following elements:

- a box, where the bottom and top are the first and third quartiles
- a red line indicating the second quartile, i.e. the median
- whiskers extending to the maximum and minimum
- red crosses indicating outliers.

The maximum is defined as the data point with the highest value still within 1.5 times the inter quartile range (i.e. the third quartile minus first quartile) of the third quartile. The minimum is defined accordingly. Data points are interpreted as outliers if located outside of the extent of the whiskers.
In Section 8.4 we use diagrams with two (color indicated) vertical axes to depict corresponding mean orientation and displacement errors. Filled boxes are used for results based on the planar motion model, whereas as empty boxes are used for results based on the general motion model.

Figure number	Number of data points out of bounds
Figure 8.6	(4,8), (6,7)
Figure 8.7a	(4,8), (3,9), (0,1), (0,13), (5,28), (22,53)
Figure 8.7b	(6,7), (1,3), (2,13), (12,33), (29,77)
Figure 8.8	(0,0), (2,5)
Figure 8.9	(4,8), (6,7), (1,1), (4,6), (0,0), (2,5)
Figure 8.11	(5,17), (6,8), (0,0)

Table A.1: Number of data points not shown in the box plots in Chapter 8.

Figure number	Median values (°, millimeters)
Figure 8.6	(0.22, 36.5), (0.17, 35.4),
Figure 8.7a	(0.22, 36.5), (0.23, 37.5), (0.21, 40.1), (0.23, 43.6), (0.25, 52.6), (0.26, 62.7)
Figure 8.7b	(0.17, 35.4), (0.19, 33.4), (0.21, 38.5), (0.21, 48.9), (0.23, 55.6)
Figure 8.8	(0.10, 16.2), (0.10, 10.8)
Figure 8.9	(0.22, 36.5), (0.17, 35.4), (0.16, 30.2), (0.13, 31.9), (0.10, 16.2), (0.10, 10.8)
Figure 8.11	(0.028, 6.72), (0.009, 4.93), (0.008, 4.00)

Table A.2: Median values for box plots shown in Chapter 8.

A.9 Additional Information on Quantitative Results

For completeness, tables A.1 and A.2 provide additional information with respect to the box plots shown in Chapter 8. Table A.1 shows the number of data points not displayed in the box plots, and Table A.2 provides the median values.

Bibliography

[1] S. Abraham and W. Förstner. Fish-eye-stereo calibration and epipolar rectification. *ISPRS Journal of Photogrammetry and Remote Sensing*, 59(5):278–288, 2005.

[2] M. Abramowitz and I. A. Stegun. *Handbook of Mathematical Functions with Formulas, Graphs, and Mathematical Tables*. Dover, New York, ninth dover printing, tenth gpo printing edition, 1964.

[3] M. Aggarwal, Hong Hua, and N. Ahuja. On cosine-fourth and vignetting effects in real lenses. In *Computer Vision, 2001. ICCV 2001. Proceedings. Eighth IEEE International Conference on*, volume 1, pages 472–479 vol.1, 2001.

[4] J. Arrospide, L. Salgado, M. Nieto, and R. Mohedano. Homography-based ground plane detection using a single on-board camera. *Intelligent Transport Systems, IET*, 4(2):149–160, 2010.

[5] F. Bajramovic and J. Denzler. Global uncertainty-based selection of relative poses for multi camera calibration. In *Proceedings of the British Machine Vision Conference*, page 74.1. BMVA Press, 2008.

[6] Y. Bar-Shalom and X.-R. Li. *Estimation and tracking: Principles, techniques, and software*. Artech House, Boston, 1993.

[7] J. Becker, M.-B. A. Colas, S. Nordbruch, and M. Fausten. *Bosch's Vision and Roadmap Toward Fully Autonomous Driving*, pages 49–59. Springer International Publishing, Cham, 2014.

[8] K. Bengler, K. Dietmayer, B. Farber, M. Maurer, C. Stiller, and H. Winner. Three decades of driver assistance systems: Review and future perspectives. *IEEE Intelligent Transportation Systems Magazine*, 6(4):6–22, 2014.

[9] J. Brookshire and S. J. Teller. Automatic calibration of multiple coplanar sensors los angeles, ca, usa, june 27-30, 2011. In Hugh F. Durrant-Whyte, Nicholas Roy, and Pieter Abbeel, editors, *Robotics: Science and Systems VII, University of Southern California, Los Angeles, CA, USA, June 27-30, 2011*, 2011.

[10] J. Brookshire and S. J. Teller. Extrinsic calibration from per-sensor ego-motion. In *Proceedings of Robotics: Science and Systems*, Sydney, Australia, 2012.

[11] M. Brown and D. Lowe. Invariant features from interest point groups. In *Proceedings of the British Machine Vision Conference*, page 23.1. BMVA Press, 2002.

[12] P. Burt, L. Wixson, and G. Salgian. Electronically directed focal stereo. In *Computer Vision, 1995. Proceedings., Fifth International Conference on*, pages 94–101, 1995.

[13] M. Calonder, V. Lepetit, C. Strecha, and P. Fua. Brief: Binary robust independent elementary features. In *Proceedings of the 11th European Conference on Computer Vision: Part IV*, ECCV'10, pages 778–792, Berlin, Heidelberg, 2010. Springer-Verlag.

[14] G. Carrera, A. Angeli, and A. J. Davison. Slam-based automatic extrinsic calibration of a multi-camera rig. In *Robotics and Automation (ICRA), 2011 IEEE International Conference on*, pages 2652–2659, 2011.

[15] Y. Caspi and M. Irani. Aligning non-overlapping sequences. *International Journal of Computer Vision*, 48(1):39–51, 2002.

[16] H. Chen, P. Meer, and D. E. Tyler. Robust regression for data with multiple structures. In *Computer Vision and Pattern Recognition, 2001. CVPR 2001. Proceedings of the 2001 IEEE Computer Society Conference on*, volume 1, pages 1069–1075, 2001.

[17] D. Comaniciu and P. Meer. Mean shift: a robust approach toward feature space analysis. *Pattern Analysis and Machine Intelligence, IEEE Transactions on*, 24(5):603–619, 2002.

[18] T. Dang, C. Hoffmann, and C. Stiller. Continuous stereo self-calibration by camera parameter tracking. *IEEE Transactions on image processing : a publication of the IEEE Signal Processing Society*, 18(7):1536–1550, 2009.

[19] S. Esquivel, F. Woelk, and R. Koch. Calibration of a multi-camera rig from non-overlapping views. In *Proceedings of the 29th DAGM Conference on Pattern Recognition*, pages 82–91, Berlin, Heidelberg, 2007. Springer-Verlag.

[20] M. A. Fischler and R. C. Bolles. Random sample consensus: A paradigm for model fitting with applications to image analysis and automated cartography. *Commun. ACM*, 24(6):381–395, 1981.

[21] M. A. Fischler and O. Firschein, editors. *Readings in Computer Vision: Issues, Problem, Principles, and Paradigms.* Morgan Kaufmann Publishers Inc, San Francisco, CA, USA, 1987.

[22] W. Fleming. Forty-year review of automotive electronics: A unique source of historical information on automotive electronics. *IEEE Vehicular Technology Magazine*, 10(3):80–90, 2015.

[23] D. B. Gennery. Generalized camera calibration including fish-eye lenses. *International Journal of Computer Vision*, 68(3):239–266, 2006.

[24] D. B. Goldman. Vignette and exposure calibration and compensation. *Pattern Analysis and Machine Intelligence, IEEE Transactions on*, 32(12):2276–2288, 2010.

[25] G. H. Golub and Van Loan, Charles F. *Matrix Computations (3rd Ed.).* Johns Hopkins University Press, Baltimore, MD, USA, 1996.

[26] S. I. Granshaw. Bundle adjustment methods in engineering photogrammetry. *The Photogrammetric Record*, 10(56):181–207, 1980.

[27] P. Greisen, M. Schaffner, S. Heinzle, M. Runo, A. Smolic, A. Burg, H. Kaeslin, and M. Gross. Analysis and vlsi implementation of ewa rendering for real-time hd video applications. *Circuits and Systems for Video Technology, IEEE Transactions on*, 22(11):1577–1589, 2012.

[28] H. M. Gross, H. Boehme, C. Schroeter, S. Mueller, A. Koenig, E. Einhorn, C. Martin, M. Merten, and A. Bley. Toomas: Interactive shopping guide robots in everyday use - final implementation and experiences from long-term field trials. In *Intelligent Robots and Systems, 2009. IROS 2009. IEEE/RSJ International Conference on*, pages 2005–2012, 2009.

[29] F. M. Ham and R. G. Brown. Observability, eigenvalues, and kalman filtering. *IEEE Transactions on Aerospace and Electronic Systems*, AES-19(2):269–273, 1983.

[30] P. Hansen, H. Alismail, P. Rander, and B. Browning. Online continuous stereo extrinsic parameter estimation. In *2012 IEEE Conference on Computer Vision and Pattern Recognition (CVPR)*, pages 1059–1066, 2012.

[31] R. Hartley and A. Zisserman. *Multiple view geometry in computer vision.* Cambridge University Press, Cambridge, UK and New York, 2nd ed edition, 2003.

[32] P. S. Heckbert. Fundamentals of texture mapping and image warping, 1989.

[33] J. Heinly, E. Dunn, and J.-M. Frahm. Comparative evaluation of binary features. In A. Fitzgibbon, S. Lazebnik, P. Perona, Y. Sato, and C. Schmid, editors, *Computer Vision – ECCV 2012*, Lecture Notes in Computer Science, pages 759–773. Springer Berlin Heidelberg, 2012.

[34] L. Heng, M. Bürki, G. H. Lee, P. Furgale, R. Siegwart, and M. Pollefeys. Infrastructure-based calibration of a multi-camera rig. In *Proceedings of the IEEE International Conference on Robotics and Automation (ICRA)*, Hong Kong, China, 2014.

[35] L. Heng, P. T. Furgale, and M. Pollefeys. Leveraging image-based localization for infrastructure-based calibration of a multi-camera rig. *J. Field Robotics*, 32(5):775–802, 2015.

[36] L. Heng, B. Li, and M. Pollefeys. Camodocal: Automatic intrinsic and extrinsic calibration of a rig with multiple generic cameras and odometry. In *Intelligent Robots and Systems (IROS), 2013 IEEE/RSJ International Conference on*, pages 1793–1800, 2013.

[37] J. Horgan, C. Hughes, J. McDonald, and S. Yogamani. Vision-based driver assistance systems: Survey, taxonomy and advances. In *Intelligent Transportation Systems (ITSC), 2015 IEEE 18th International Conference on*, pages 2032–2039, 2015.

[38] B. Jähne. *Practical Handbook on Image Processing for Scientific and Technical Applications*. CRC Press, 2 edition, 2004.

[39] S. J. Julier and J. J. LaViola. On kalman filtering with nonlinear equality constraints. *Signal Processing, IEEE Transactions on*, 55(6):2774–2784, 2007.

[40] M. Kaess, A. Ranganathan, and F. Dellaert. isam: Incremental smoothing and mapping. *IEEE Transactions on Robotics*, 24(6):1365–1378, 2008.

[41] T. Kazik, L. Kneip, J. Nikolic, M. Pollefeys, and R. Siegwart. Real-time 6d stereo visual odometry with non-overlapping fields of view. In *Computer Vision and Pattern Recognition (CVPR), 2012 IEEE Conference on*, pages 1529–1536, 2012.

[42] J. Klappstein, F. Stein, and U. Franke. Applying kalman filtering to road homography estimation. In *Workshop on Planning, Perception and Navigation for Intelligent Vehicles held at IEEE International Conference on Robotics and Automation (ICRA)*, 2007.

[43] M. Knorr, J. Esparza, W. Niehsen, and C. Stiller. Extrinsic calibration of a fisheye multi-camera setup using overlapping fields of view. In *Intelligent Vehicles Symposium (IV), 2014 IEEE*, pages 1276–1281, 2014.

[44] M. Knorr, W. Niehsen, and C. Stiller. Online extrinsic multi-camera calibration using ground plane induced homographies. In *Intelligent Vehicles Symposium (IV), 2013 IEEE*, pages 236–241, 2013.

[45] M. Knorr, W. Niehsen, and C. Stiller. Robust ground plane induced homography estimation for wide angle fisheye cameras. In *Intelligent Vehicles Symposium (IV), 2014 IEEE*, pages 1288–1293, 2014.

[46] R. Kuemmerle, G. Grisetti, H. Strasdat, K. Konolige, and W. Burgard. g2o: A general framework for graph optimization. In *Proceedings of the IEEE International Conference on Robotics and Automation (ICRA)*, pages 3607–3613, Shanghai, China, 2011.

[47] R. Kumar, P. Anandan, and K. Hanna. Shape recovery from multiple views: a parallax based approach. In *DARPA Image Understanding Workshop*, 1994.

[48] R. K. Kumar, A. Ilie, J.-M. Frahm, and M. Pollefeys. Simple calibration of non-overlapping cameras with a mirror. In *Computer Vision and Pattern Recognition, 2008. CVPR 2008. IEEE Conference on*, pages 1–7, 2008.

[49] B. Lamprecht, S. Rass, S. Fuchs, and K. Kyamakya. Extrinsic camera calibration for an on-board two-camera system without overlapping field of view. In *Intelligent Transportation Systems Conference, 2007. ITSC 2007. IEEE*, pages 265–270, 2007.

[50] P. Lébraly, C. Deymier, O. Ait-Aider, E. Royer, and M. Dhome. Flexible extrinsic calibration of non-overlapping cameras using a planar mirror: Application to vision-based robotics. In *Intelligent Robots and Systems (IROS), 2010 IEEE/RSJ International Conference on*, pages 5640–5647, 2010.

[51] P. Lébraly, E. Royer, O. Ait-Aider, and M. Dhome. Calibration of non-overlapping cameras - application to vision-based robotics. In *Proceedings of the British Machine Vision Conference*, page 10.1. BMVA Press, 2010.

[52] S. Leutenegger, M. Chli, and R. Y. Siegwart. Brisk: Binary robust invariant scalable keypoints. In *Computer Vision (ICCV), 2011 IEEE International Conference on*, pages 2548–2555, 2011.

[53] B. Li, L. Heng, K. Köser, and M. Pollefeys. A multiple-camera system calibration toolbox using a feature descriptor-based calibration pattern. In *Intelligent Robots and Systems (IROS), 2013 IEEE/RSJ International Conference on*, pages 1301–1307, 2013.

[54] S. Li and Y. Hai. Easy calibration of a blind-spot-free fisheye camera system using a scene of a parking space. *IEEE Transactions on Intelligent Transportation Systems*, 12(1):232–242, 2011.

[55] M. I. A. Lourakis and A. A. Argyros. Chaining planar homographies for fast and reliable 3d plane tracking. In *Pattern Recognition, 2006. ICPR 2006. 18th International Conference on*, volume 1, pages 582–586, 2006.

[56] D. G. Lowe. Object recognition from local scale-invariant features. In *Computer Vision, 1999. The Proceedings of the Seventh IEEE International Conference on*, volume 2, pages 1150–1157 vol.2, 1999.

[57] T. Luhmann. *Close range photogrammetry: Principles, methods and applications*. Whittles, Dunbeath, 2006.

[58] Q.-T. Luong and O. D. Faugeras. Self-calibration of a stereo rig from unknown camera motions and point correspondences. In A. Gruen and T. S. Huang, editors, *Calibration and Orientation of Cameras in Computer Vision*, pages 195–229. Springer Berlin Heidelberg, Berlin, Heidelberg, 2001.

[59] E. Malis and M. Vargas. Deeper understanding of the homography decomposition for vision-based control: Research report, 2007.

[60] F. L. Markley. Attitude error representations for kalman filtering. *Journal of Guidance, Control, and Dynamics*, 26(2):311–317, 2003.

[61] S. J. Maybank. Filter based estimates of depth. In *Proceedings of the British Machine Vision Conference*, pages 62.1–62.6. BMVA Press, 1990.

[62] J. Maye. *Online Self-Calibration for Robotic Systems*. PhD thesis, Eidgenössische Technische Hochschule ETH Zürich, 2014.

[63] J. Maye, P. Furgale, and R. Siegwart. Self-supervised calibration for robotic systems. In *Intelligent Vehicles Symposium (IV), 2013 IEEE*, pages 473–480, 2013.

[64] C. Mei and P. Rives. Single view point omnidirectional camera calibration from planar grids. In *IEEE International Conference on Robotics and Automation (ICRA)*, pages 3945–3950, Rome, Italie, 2007. IEEE.

[65] K. Mikolajczyk, T. Tuytelaars, C. Schmid, A. Zisserman, J. Matas, F. Schaffalitzky, T. Kadir, and L. Gool. A comparison of affine region detectors. *International Journal of Computer Vision*, 65(1-2):43–72, 2005.

[66] M. Miksch, Bin Yang, and K. Zimmermann. Automatic extrinsic camera self-calibration based on homography and epipolar geometry. In *Intelligent Vehicles Symposium (IV), 2010 IEEE*, pages 832–839, 2010.

[67] M. Miksch, B. Yang, and K. Zimmermann, editors. *Homography-based extrinsic self-calibration for cameras in automotive applications*, 2010.

[68] J.-M. Morel and G. Yu. Asift: A new framework for fully affine invariant image comparison. *SIAM J. Img. Sci.*, 2(2):438–469, 2009.

[69] G. R. Mueller and H.-J. Wuensche. Continuous extrinsic online calibration for stereo cameras. In *Intelligent Vehicles Symposium (IV), 2016 IEEE*, pages 1047–1052, 2016.

[70] D. Muhle, S. Abraham, C. Heipke, and M. Wiggenhagen. Estimating the mutual orientation in a multi-camera system with a non overlapping field of view. In U. Stilla, F. Rottensteiner, H. Mayer, B. Jutzi, and M. Butenuth, editors, *Photogrammetric Image Analysis: ISPRS Conference, PIA 2011, Munich, Germany, October 5-7, 2011. Proceedings*, pages 13–24. Springer Berlin Heidelberg, Berlin, Heidelberg, 2011.

[71] D. Nistér. An efficient solution to the five-point relative pose problem. *Pattern Analysis and Machine Intelligence, IEEE Transactions on*, 26(6):756–770, 2004.

[72] D. Nistér, O. Naroditsky, and J. Bergen. Visual odometry. In *Computer Vision and Pattern Recognition, 2004. CVPR 2004. Proceedings of the 2004 IEEE Computer Society Conference on*, volume 1, pages I–652–I–659 Vol.1, 2004.

[73] F. Pagel. *Kalibrierung mobiler Multikamerasysteme mit disjunkten Sichtfeldern.* Dissertation, Karlsruher Institut für Technologie, Karlsruhe, 2012.

[74] F. Pagel. Motion adjustment for extrinsic calibration of cameras with non-overlapping views. In *Computer and Robot Vision (CRV), 2012 Ninth Conference on*, pages 94–100, 2012.

[75] F. Pagel. Extrinsic self-calibration of multiple cameras with non-overlapping views in vehicles. *Proc. SPIE*, 9026, 2014.

[76] F. Pagel and D. Willersinn, editors. *Extrinsic Camera Calibration in Vehicles with Explicit Ground Estimation*, 2011.

[77] B. Ranft and C. Stiller. The role of machine vision for intelligent vehicles. *IEEE Transactions on Intelligent Vehicles*, 1(1):8–19, 2016.

[78] Reinhard Diestel. *Graph Theory, 4th Edition*, volume 173 of *Graduate texts in mathematics*. Springer, 2012.

[79] E. Rosten and T. Drummond. Machine learning for high-speed corner detection. In A. Leonardis, H. Bischof, and A. Pinz, editors, *Computer Vision – ECCV 2006*, volume 3951 of *Lecture Notes in Computer Science*, pages 430–443. Springer Berlin Heidelberg, 2006.

[80] E. Rublee, V. Rabaud, K. Konolige, and G. Bradski. Orb: An efficient alternative to sift or surf. In *Proceedings of the 2011 International Conference on Computer Vision*, ICCV '11, pages 2564–2571, Washington, DC, USA, 2011. IEEE Computer Society.

[81] T. Ruland, H. Loose, T. Pajdla, and L. Krüger. Hand-eye autocalibration of camera positions on vehicles. In *Intelligent Transportation Systems (ITSC), 2010 13th International IEEE Conference on*, pages 367–372, 2010.

[82] T. Ruland, T. Pajdla, and L. Krüger. Extrinsic autocalibration of vehicle mounted cameras for maneuvering assistance. In *Proceedings of the Computer Vision Winter Workshop*, pages 44–51. 2010.

[83] H. S. Sawhney. 3d geometry from planar parallax. In *Computer Vision and Pattern Recognition, 1994. Proceedings CVPR '94., 1994 IEEE Computer Society Conference on*, pages 929–934, 1994.

[84] D. Scaramuzza. 1-point-ransac structure from motion for vehicle-mounted cameras by exploiting non-holonomic constraints. *International Journal of Computer Vision*, 95(1):74–85, 2011.

[85] S. Schneider, T. Luettel, and H. J. Wuensche. Odometry-based online extrinsic sensor calibration. In *2013 IEEE/RSJ International Conference on Intelligent Robots and Systems*, pages 1287–1292, 2013.

[86] M. Schoenbein, T. Strauss, and A. Geiger. Calibrating and centering quasi-central catadioptric cameras. In *International Conference on Robotics and Automation (ICRA)*, 2014.

[87] Y. C. Shiu and S. Ahmad. Calibration of wrist-mounted robotic sensors by solving homogeneous transform equations of the form ax=xb. *Robotics and Automation, IEEE Transactions on*, 5(1):16–29, 1989.

[88] A. R. Smith. Digital filtering tutorial for computer graphics, 1983.

[89] M. Stämpfle. Surround sensing for automotive driver assistance systems: Communication in transportation systems. In O. Strobel, editor, *Communication in transportation systems*, Premier reference source. Information Science Reference, Hershey PA, 2013.

[90] G. P. Stein, O. Mano, and A. Shashua. A robust method for computing vehicle ego-motion. In *IV 2000 Intelligent Vehicles Symposium*, pages 362–368, 3-5 Oct. 2000.

[91] C. V. Stewart. Bias in robust estimation caused by discontinuities and multiple structures. *Pattern Analysis and Machine Intelligence, IEEE Transactions on*, 19(8):818–833, 1997.

[92] C. Stiller. Towards intelligent automotive vision systems. In L. Vlacic, M. Parent, and F. Harashima, editors, *Intelligent Vehicle Technologies: Theory and Applications*, Automotive engineering, pages 113–130. SAE International, 2001.

[93] T. Strauss, J. Ziegler, and J. Beck. Calibrating multiple cameras with non-overlapping views using coded checkerboard targets. In *Intelligent Transportation Systems (ITSC), 2014 IEEE 17th International Conference on*, pages 2623–2628, 2014.

[94] P. Sturm, S. Ramalingam, J.-P. Tardif, S. Gasparini, and J. Barreto. Camera models and fundamental concepts used in geometric computer vision. *Found. Trends. Comput. Graph. Vis.*, 6(1–2):1–183, 2011.

[95] R. Szeliski, S. Winder, and M. Uyttendaele. High-quality multi-pass image resampling, 2010.

[96] R. Toldo and A. Fusiello. Robust multiple structures estimation with j-linkage. In *Proceedings of the 10th European Conference on Computer Vision: Part I*, ECCV '08, pages 537–547, Berlin, Heidelberg, 2008. Springer-Verlag.

[97] B. Triggs. Plane + parallax, tensors and factorization. In *Computer Vision - ECCV 2000*, volume 1842 of *Lecture Notes in Computer Science*, pages 522–538. Springer Berlin Heidelberg, 2000.

[98] B. Triggs, P. McLauchlan, R. Hartley, and A. Fitzgibbon. Bundle adjustment — a modern synthesis. In B. Triggs, A. Zisserman, and R. Szeliski, editors, *Vision Algorithms: Theory and Practice*, volume 1883 of *Lecture Notes in Computer Science*, pages 298–372. Springer Berlin Heidelberg, 2000.

[99] R. Y. Tsai and R. K. Lenz. Real time versatile robotics hand/eye calibration using 3d machine vision. In *Robotics and Automation, 1988. Proceedings., 1988 IEEE International Conference on*, pages 554–561 vol.1, 1988.

[100] R. Y. Tsai and R. K. Lenz. A new technique for fully autonomous and efficient 3d robotics hand/eye calibration. *Robotics and Automation, IEEE Transactions on*, 5(3):345–358, 1989.

[101] J. W. Tukey. *Exploratory Data Analysis*. Behavioral Science: Quantitative Methods. Addison-Wesley, Reading, Mass., 1977.

[102] S. Urban and S. Hinz. mdbrief - a fast online adaptable, distorted binary descriptor for real-time applications using calibrated wide-angle or fisheye cameras. *CoRR*, abs/1610.07804, 2016.

[103] S. Urban, S. Wursthorn, J. Leitloff, and S. Hinz. Multicol bundle adjustment: A generic method for pose estimation, simultaneous self-calibration and reconstruction for arbitrary multi-camera systems. *International Journal of Computer Vision*, pages 1–19, 2016.

[104] K. Yamaguchi, T. Kato, and Y. Ninomiya. Vehicle ego-motion estimation and moving object detection using a monocular camera. In *Pattern Recognition, 2006. ICPR 2006. 18th International Conference on*, volume 4, pages 610–613, 2006.

[105] Z. Zhang. A flexible new technique for camera calibration. *IEEE Trans. Pattern Anal. Mach. Intell.*, 22(11):1330–1334, Nov. 2000.

Schriftenreihe
Institut für Mess- und Regelungstechnik
Karlsruher Institut für Technologie
(1613-4214)

Band 001 Hans, Annegret
Entwicklung eines Inline-Viskosimeters auf Basis eines magnetisch-induktiven Durchflussmessers. 2004
ISBN 3-937300-02-3

Band 002 Heizmann, Michael
Auswertung von forensischen Riefenspuren mittels automatischer Sichtprüfung. 2004
ISBN 3-937300-05-8

Band 003 Herbst, Jürgen
Zerstörungsfreie Prüfung von Abwasserkanälen mit Klopfschall. 2004
ISBN 3-937300-23-6

Band 004 Kammel, Sören
Deflektometrische Untersuchung spiegelnd reflektierender Freiformflächen. 2005
ISBN 3-937300-28-7

Band 005 Geistler, Alexander
Bordautonome Ortung von Schienenfahrzeugen mit Wirbelstrom-Sensoren. 2007
ISBN 978-3-86644-123-1

Band 006 Horn, Jan
Zweidimensionale Geschwindigkeitsmessung texturierter Oberflächen mit flächenhaften bildgebenden Sensoren. 2007
ISBN 978-3-86644-076-0

Die Bände sind unter www.ksp.kit.edu als PDF frei verfügbar oder als Druckausgabe bestellbar.

Band 007 Hoffmann, Christian
Fahrzeugdetektion durch Fusion monoskopischer Videomerkmale. 2007
ISBN 978-3-86644-139-2

Band 008 Dang, Thao
Kontinuierliche Selbstkalibrierung von Stereokameras. 2007
ISBN 978-3-86644-164-4

Band 009 Kapp, Andreas
Ein Beitrag zur Verbesserung und Erweiterung der Lidar-Signalverarbeitung für Fahrzeuge. 2007
ISBN 978-3-86644-174-3

Band 010 Horbach, Jan
Verfahren zur optischen 3D-Vermessung spiegelnder Oberflächen. 2008
ISBN 978-3-86644-202-3

Band 011 Böhringer, Frank
Gleisselektive Ortung von Schienenfahrzeugen mit bordautonomer Sensorik. 2008
ISBN 978-3-86644-196-5

Band 012 Xin, Binjian
Auswertung und Charakterisierung dreidimensionaler Messdaten technischer Oberflächen mit Riefentexturen. 2009
ISBN 978-3-86644-326-6

Band 013 Cech, Markus
Fahrspurschätzung aus monokularen Bildfolgen für innerstädtische Fahrerassistenzanwendungen. 2009
ISBN 978-3-86644-351-8

Band 014 Speck, Christoph
Automatisierte Auswertung forensischer Spuren auf Patronenhülsen. 2009
ISBN 978-3-86644-365-5

Band 015	Bachmann, Alexander **Dichte Objektsegmentierung in Stereobildfolgen.** 2010 ISBN 978-3-86644-541-3
Band 016	Duchow, Christian **Videobasierte Wahrnehmung markierter Kreuzungen mit lokalem Markierungstest und Bayes'scher Modellierung.** 2011 ISBN 978-3-86644-630-4
Band 017	Pink, Oliver **Bildbasierte Selbstlokalisierung von Straßenfahrzeugen.** 2011 ISBN 978-3-86644-708-0
Band 018	Hensel, Stefan **Wirbelstromsensorbasierte Lokalisierung von Schienenfahrzeugen in topologischen Karten.** 2011 ISBN 978-3-86644-749-3
Band 019	Carsten Hasberg **Simultane Lokalisierung und Kartierung spurgeführter Systeme.** 2012 ISBN 978-3-86644-831-5
Band 020	Pitzer, Benjamin **Automatic Reconstruction of Textured 3D Models.** 2012 ISBN 978-3-86644-805-6
Band 021	Roser, Martin **Modellbasierte und positionsgenaue Erkennung von Regentropfen in Bildfolgen zur Verbesserung von videobasierten Fahrerassistenzfunktionen.** 2012 ISBN 978-3-86644-926-8

Band 022 Loose, Heidi
Dreidimensionale Straßenmodelle für Fahrerassistenzsysteme auf Landstraßen. 2013
ISBN 978-3-86644-942-8

Band 023 Rapp, Holger
Reconstruction of Specular Reflective Surfaces using Auto-Calibrating Deflectometry. 2013
ISBN 978-3-86644-966-4

Band 024 Moosmann, Frank
Interlacing Self-Localization, Moving Object Tracking and Mapping for 3D Range Sensors. 2013
ISBN 978-3-86644-977-0

Band 025 Geiger, Andreas
Probabilistic Models for 3D Urban Scene Understanding from Movable Platforms. 2013
ISBN 978-3-7315-0081-0

Band 026 Hörter, Marko
Entwicklung und vergleichende Bewertung einer bildbasierten Markierungslichtsteuerung für Kraftfahrzeuge. 2013
ISBN 978-3-7315-0091-9

Band 027 Kitt, Bernd
Effiziente Schätzung dichter Bewegungsvektorfelder unter Berücksichtigung der Epipolargeometrie zwischen unterschiedlichen Ansichten einer Szene. 2013
ISBN 978-3-7315-0105-3

Band 028 Lategahn, Henning
Mapping and Localization in Urban Environments Using Cameras. 2013
ISBN 978-3-7315-0135-0

Band 029 Tischler, Karin
Informationsfusion für die kooperative Umfeldwahrnehmung vernetzter Fahrzeuge. 2014
ISBN 978-3-7315-0166-4

Band 030 Schmidt, Christian
Fahrstrategien zur Unfallvermeidung im Straßenverkehr für Einzel- und Mehrobjektszenarien. 2014
ISBN 978-3-7315-0198-5

Band 031 Firl, Jonas
Probabilistic Maneuver Recognition in Traffic Scenarios. 2014
ISBN 978-3-7315-0287-6

Band 032 Schönbein, Miriam
Omnidirectional Stereo Vision for Autonomous Vehicles. 2015
ISBN 978-3-7315-0357-6

Band 033 Nicht erschienen

Band 034 Liebner, Martin
Fahrerabsichtserkennung und Risikobewertung für warnende Fahrerassistenzsysteme. 2016
ISBN 978-3-7315-0508-2

Band 035 Ziegler, Julius
Optimale Trajektorienplanung für Automobile. 2017
ISBN 978-3-7315-0553-2

Band 036 Harms, Hannes
Genauigkeitsuntersuchung von binokularen Normalenvektoren für die Umfeldwahrnehmung. 2017
ISBN 978-3-7315-0628-7

Band 037 Ruhhammer, Christian
Inferenz von Kreuzungsinformationen aus Flottendaten. 2017
ISBN 978-3-7315-0721-5

Band 038 Stein, Denis
Mobile laser scanning based determination of railway network topology and branching direction on turnouts. 2018
ISBN 978-3-7315-0743-7

Band 039 Yi, Boliang
Integrated Planning and Control for Collision Avoidance Systems. 2018
ISBN 978-3-7315-0785-7

Band 040 Schwarze, Tobias
Compact Environment Modelling from Unconstrained Camera Platforms. 2018
ISBN 978-3-7315-0801-4

Band 041 Knorr, Moritz
Self-Calibration of Multi-Camera Systems for Vehicle Surround Sensing. 2018
ISBN 978-3-7315-0765-9